BARRY'S CONSTRUCTION TECHNOLOGY

BARRY'S CONSTRUCTION TECHNOLOGY

Analysis and Choice

Third Edition

Stephen Emmitt
University of Bath
United Kingdom

Tony Bryan
United Kingdom

WILEY Blackwell

Edition History
Wiley Blackwell (1e, 2005), Wiley Blackwell (2e, 2010).

Registered Offices
John Wiley & Sons, Inc., 111 River Street, Hoboken, NJ 07030, USA
John Wiley & Sons Ltd, New Era House, 8 Oldlands Way, Bognor Regis, West Sussex, PO22 9NQ, UK

For details of our global editorial offices, customer services, and more information about Wiley products visit us at www.wiley.com.

The manufacturer's authorized representative according to the EU General Product Safety Regulation is Wiley-VCH GmbH, Boschstr. 12, 69469 Weinheim, Germany, e-mail: Product_Safety@wiley.com.

Wiley also publishes its books in a variety of electronic formats and by print-on-demand. Some content that appears in standard print versions of this book may not be available in other formats.

Library of Congress Cataloging-in-Publication Data Applied for:

Paperback ISBN: 9781394264643

Cover Design: Wiley
Cover Image: © Jason marz/Getty Images

Set in size of 10/12pt and Minion Pro by Straive, Chennai, India
Printed and bound by CPI Group (UK) Ltd, Croydon, CR0 4YY
C9781394264643_070825

Contents

Preface

This book has been written for readers at the beginning of their career, studying with a view to practise in one of the professions concerned with the design, engineering, construction, upgrading and recycling of buildings. It is written for the novice who, with study and experience, will become an expert in their chosen field. The more fundamental ideas developed in the text may need to be revisited as experience grows, and the book can be referred to in the early years of practice and maybe beyond.

The book takes as its theme the process of choice: what the expert must know and how the expert could think through the myriad of decisions that have to be made about the design, production, maintenance, adaptation, reuse, and recycling of buildings. While this involves a process of analysis, the basis of which is universally applicable, the final choice is dictated by context. The context is set by time and place, available finance, the nature of the activity to be undertaken in the building and the aspirations of its benefactor, which includes the client, the building users and more widely the local community. The process of choice is about seeking appropriate solutions in a physical and cultural context. This decision-making process is made easier in some respects by the uptake of digital tools that help to provide 'the solution' to complex challenges; and made harder in other respects because the reasoning built into the software is often hidden from the decision maker. Regardless of the level of digital sophistication used to assist analysis and choice, there is still a need for professionals to understand the fundamentals, and hence be equipped to challenge automated solutions. Relying, for example, on standard solutions contained in Building Information Modelling (BIM) libraries, which are often copied without sufficient scrutiny and questioning, is not what a client is paying for. Clients employ professionals to make informed decisions, drawing on their expertise and experience to deliver value. This is enhanced by using digital tools and technologies to help with analysis and choice.

For a text of this nature it may seem to be dominated by words and not illustrations. The contention is that for the novice or the practising professional faced with new challenges, ideas, and situations, an explanation of possible solutions is required. Presenting a final working drawing may be the outcome of choice, but it is the process of analysis, with its need for explanations, that gives the confidence that the solution will perform as intended, and it can be built safely and economically. A picture only paints a thousand words if you already have the words to explain it. We hope that the reader will find that the text tells a good story, that the development of the ideas is logical and easy to follow, but perhaps above all that it is seen as valuable in the world of modern construction practice. It is also hoped that the text will be illuminating and the illustrations supportive in providing the explanations of both the analysis and the solutions suggested for choice.

This third edition of the book has been reimagined to be part of the Barry's Construction of Buildings series, and readers are directed to *Barry's Introduction to Construction of Buildings* and *Barry's Advanced Construction of Buildings* for guidance on typical solutions to familiar construction challenges. This has allowed for simplification of content and a strengthening of the analysis and choice philosophy within the text. We have strengthened the sustainability and resilience argument by considering the circular construction economy and buildings as resource bases. We have also moved the chapter on sustainability to earlier in Part One of the book and refocused the second part of the book to making more sustainable and resilient choices. We have simplified and condensed the second part of the book, concentrating on a small number of building elements to challenge the reader. Rather than repeat many of the typical details in the Barry books, we have tried to suggest ways forward by challenging readers to think creatively and objectively.

The first part of the book focuses on the key areas of analysis that must be considered to check that the proposed construction will perform (will not fail), can be safely built, and performs as intended during its service life. These tests are seen as fundamental to the final choice. To carry out these analyses this book suggests that it is necessary to visualise not only the physical form of the final construction, however outline or tentative, but also, equally importantly, its response to the dynamic conditions under which it must perform. The key question is 'What if I do it like this?' The answer comes from an understanding of how it works and how it is built, this understanding coming from explanations of why things happen, drawing on knowledge from several other disciplines, such as science and economics. Each of the areas of analysis presented in this book is independent, and therefore each analysis must be separate, yet all must be satisfactory before the final choice is made. Choice is a juggling act: checking a proposal in one area of analysis may suggest changes that will affect others. For example, structurally sound solutions may be difficult to build, or solutions that will last without maintenance may be too expensive as a capital investment. There will, therefore, need to be some compromise to achieve a balance of performance criteria in the final choice that satisfies all areas of analysis.

The second part of the book puts analysis into practice, focusing on making a choice for a thought-provoking situation. The chapters are designed to challenge the reader and go beyond conventional solutions to seek original solutions to the challenges of our time. One of the drivers behind these challenges is the climate emergency. Another driver is the need to provide a human centric built environment that benefits all members of society. Both drivers should inform analysis and choice.

The background to the book content is grounded in the regulations relating to the UK, however, the scenarios are deliberately designed to appeal to readers with different contexts and climates, as the principles are widely applicable. The notion of the framework for understanding developed in the book encourages the idea of seeking relevant and current information in the context of each project. Following this argument, the book does not give references to other publications, other than the Barry series. The intention is to encourage readers to conduct their own research and reflection and develop their analytical and decision-making competences. This may well link to new and emerging technologies, such as artificial intelligence (AI), machine learning (ML), and bespoke software that collectively can guide the student and practitioner towards making better informed decisions. In practice much of the workflow now follows a digital protocol, and this is starting to be reflected in education. Thus what is presented in this book should be seen as supporting and encouraging creative thinking, regardless of the digital tools to be used.

This book is written with the conviction that by focusing on the process of choice the range of theory and knowledge that is useful to practice becomes explicit, making the link between knowledge and practice, and between understanding and experience, clear. As study, and then practice, unfolds for everyone it is the power to reflect that develops understanding and continual improvement. It is hoped that the framework suggested in this book will help not just the process of choice but also the process of reflection. Readers should take from this text what understanding they can to use it to develop their own good judgement in whatever part they play in the great enterprise of construction and the development of a resilient built environment.

PART I Analysis

The Framework for Understanding: Chapter 1 AT A GLANCE

What? With the need to respond to the climate emergency and design and engineer buildings that provide a positive impact on the planet there comes a need to better understand the decisions that we make. This requires a framework for understanding. The process of choice is one of suggestions and evaluation that requires knowledge of the physical construction and the dynamic systems in which the building will be built and will operate over a long period.

Why? We need to rethink how, why, what and when we build; and where we build. When there is a changing demand for buildings and rapidly developing technical knowledge, the role of experience changes from predominantly reproducing known solutions to the integration of experience into a process of analysis to explore new possibilities. Climate change is a disrupter and the response to this disruption is to innovate. The initial suggestion will probably be based on precedent and general building forms to start the process of analysis and choice to meet current requirements, from which specific, more sustainable and resilient solutions can be derived.

When? Decisions need to be made throughout the life of a construction project and beyond through the long life of the building in service, and its eventual disassembly and reuse of precious resources into new constructions. The design and the resource availability define the criteria for the evaluation. This must be set within a physical and cultural context. How a suggestion is generated and the extent to which a formal evaluation must be undertaken will depend on the scale and nature of any changes from current practice. This will be influenced by the wants and needs of the building sponsor and the resources available at the time.

How? The majority of decisions are made collectively. Thus, everyone contributing to a building project must appreciate and understand the need for collective decision making, which requires a degree of compromise and determination to see solutions realised in a constructed artefact. Similarly, everyone contributing to the use and management of a building must appreciate the impact that their choices are having on the serviceability and usability of the building. By better understanding our collective decision making we will be better positioned to enable a carbon neutral and healthy built environment. The evaluation needs to be carried out through seven areas of analysis: appearance, behaviour creating environments, behaviour under load, behaviour over time, manufacture and assembly, cost, environmental and social concerns. This will involve constant reference to regulations, design knowledge, production expertise and maintenance intelligence.

1 The Framework for Understanding

This opening chapter outlines a framework to help develop an understanding of what must be known to take decisions on how and why we build. This is developed against a background of climate change and the impact that the built environment has on the natural environment. Responding to climate change requires a rethink about how we make decisions. This emphasises the importance of analysis and choice. The framework developed in this book suggests a way of going about selecting construction and identifies the knowledge that is required to make informed choices. It is the framework that will be developed and used throughout the book to help readers to reflect on the reasons behind the choices they make. These choices need to consider a wide variety of parameters that are underscored by the impact of the solution on the natural environment.

1.1 Responding to the climate emergency – rethinking how and why we build

It is necessary to start with the challenge facing all societies and industries and the urgent need to tackle climate change. The built environment makes a significant impact on the natural world. It is estimated that approximately 11% of greenhouse gas emissions come from the built environment. Some of the emissions are produced during the production of the building, which includes the extraction and processing of materials and the assembly of materials to realise a building project. This includes new buildings and works to the existing building stock. Some of the emissions are produced while the building is being used, primarily by using energy to create a comfortable indoor environment for the occupants. Much of the energy will be used for heating and cooling, with lighting and equipment also requiring energy. As the building weathers there will be the need for maintenance and repair, along with replacement of items that have failed or are no longer functional. As building users' needs change there may also be the need to carry out major alterations and upgrades. Sometimes this is essential to improve the serviceability of the building, sometimes it is linked to changing tastes and fashions. And when the building has finally reached the end of its service life, it provides a material bank for new building projects, with valuable materials and components recovered and reused in new building projects or repurposed into new (building) products. This prevents demolition and construction waste going to landfill and the loss of valuable resources, such as metals, bricks, concrete, glass and plastics. All these

Barry's Construction Technology: Analysis and Choice, Third Edition. Stephen Emmitt and Tony Bryan.

stages have an impact on carbon emissions and pollution. It is, therefore, imperative that as a society we start to design, build and operate buildings that make a positive impact on our natural environment, and which are resilient to a changing climate. This means that we need to reduce the carbon footprint of all construction activities and reduce the energy consumed when the building is in use. The choices that everyone makes who are involved in the design, engineering, construction, use, maintenance, upgrading and recycling of buildings will have an impact on the building's performance and its environmental impact. Thus the building needs to be envisaged in its entire life cycle to enable a more resilient outcome from the analysis and choice.

In many circumstances the 'best' and 'most appropriate' technology for a given context is not necessarily more complex or technologically more sophisticated, but a 'simpler' technology that draws on our heritage of vernacular building materials and techniques. Research and testing have shown that many 'natural' materials are a more practical choice compared to the more industrialised and processed carbon intensive materials. We have also developed a better understanding of the importance of existing buildings and the opportunities to renovate, upgrade, retrofit, disassemble and reuse. This reflects the development of a more conscientious approach in the procurement of usable space by those who commission and regulate the built environment.

Environmental consciousness is a significant driver in consumer decisions and hence a core customer value, which shapes purchasing and procurement decisions. Environmental consciousness shapes the analysis and choice. There are several ways of making a positive difference to our natural and built environment. The main argument is to reduce, reuse, recycle, regenerate and revitalise. These are themes that are developed further in this text, sometimes implicitly and sometimes explicitly. Taking this thinking forward we can view a building as a circular or closed system, where there is no waste produced during the entire life cycle of the building. This requires analysis and choice in a framework that considers the long-term future of the building, and which also requires a degree of speculation as to how the future may look. This demands a rethink about how, when, why and what we build; and where. It also requires a fundamental questioning of current building legislation, codes and design guidance to go beyond what we currently know and to innovate. This, of course, raises the spectre of risk, and the balance required between exposure to risk and the realisation of value for society. It also raises questions about building technologies. Do we put our faith in offsite production, modern methods of construction, 'greener' industrial processes, and innovative materials? Do we revisit sustainable techniques from the past and adapt them to our current expectations, relying entirely on bio-based (renewable) materials? Or is it a mixture of high-tech and low-tech choices, for example 3D printing of building components using sustainable (bio-based) materials? Does the solution have to be complex? Why cannot we return to simpler construction assemblages that embrace passive design principles, and which may be more durable in the longer term? Can we put our faith in digital technologies and software without necessarily understanding the algorithms and data that inform the solutions?

These and similar questions are easy to raise. They are less easy to answer when faced with often competing performance criteria. Fortunately, there are established design guides and green building rating systems such as BREEAM (UK), LEED (US), Green Star (Australia) and many others that help to guide professionals towards sustainable analysis and choice. These decision-making guides are complimented by a rapidly expanding array of digital tools that help with the modelling, calculation and visualisation of buildings at all stages in their lifespan. Some of these are best suited to familiar building forms and technologies, although they are becoming more sophisticated and applicable to a wider range of building forms and technologies.

Digital dilemmas and opportunities

The digital revolution has certainly made it easier to design, detail and specify a building. Building information models (BIMs) contain typologies of details that can be easily imported from the BIM library into the design, requiring little in the way of thinking or reflection. These are solutions that are known to work, based on previous experience. The problem with taking an 'easy' approach is that professionals may stop thinking and reflecting on what is required for a specific context, and more importantly what is required to tackle climate change. Standard solutions do not suit all circumstances, and one could argue that standard solutions are no longer appropriate for reducing the built environment's carbon footprint. This may be further compounded when using software to simulate and predict building performance. Do we really understand the background information that informs the software and the algorithms being used to give 'the' answer? This is not to say that digital tools are not useful; they are. The point is that professionals need to understand exactly what it is that the digital tools are contributing to the decision-making process, and how they can assist with analysis and choice. Professionals need to have the knowledge and insight to be able to question the digital 'solution' to ensure that it is a sensible choice for the context. Thus, to understand what digital tools can contribute, and how, requires an understanding of construction technology and an ability to ask the difficult questions. Too much reliance on established typologies and details may hinder the urgent need for change to a carbon free built environment. And this brings us onto the topic of artificial intelligence (AI), machine learning (ML) and generative design.

We now have the computing power and know how to generate millions of virtual solutions for a given challenge. We can model the building using BIM and virtual digital twins to examine the way in which buildings may behave if the conditions are adjusted. BIM allows the virtual construction and disassembly of the building long before the final decision is taken to go into production. This helps to reduce errors and improves coordination, as well as aiding the safe and economic construction of the building virtually before the design enters production. Virtual digital twins can be linked via the internet to sensors in buildings that gather real time data on, for example, temperature, humidity, lighting and occupancy levels, etc. This allows a reactive adjustment of building systems by using automation; known as smart or intelligent buildings. The data will inform predictive maintenance, allowing for targeted and timely interventions to maintain building performance. These data also help designers and building managers to take the learning into new, yet unknown, situations. This process is being made much easier using AI to analyse the data. AI can also contribute to a greener environment, for example by informing carbon accounting.

Digital tools are enablers of innovation and a powerful driver in the realisation of a net zero built environment. They assist decision making and generate solutions that would be impossible to do manually. The future may be an automated technological building that co-exists with a simpler form of user-control over one's immediate environment. One relies on the technology performing, for example the sensors, switches and gears opening the window; and one relies on the user performing, for example remembering to open and close the window. The argument in this book is not for one or the other, as both complexity and simplicity can, and do, co-exist.

1.2 Process and knowledge

This book provides examples of construction, showing how we currently build, and introduces the understanding necessary to explain how the decisions we make influence the performance of the completed building project. These two types of knowledge are both vital to making decisions as to how we should create and maintain buildings in the future. This book sets this knowledge in the context of the process of making choices for the construction. In practice, a great deal of knowledge exists of the types of construction we may use and the materials and details that may be specified. There is, however, the need for any proposed construction works to answer the questions 'Will it fail?' and 'Can it be built?' Clearly it requires the answer to these two questions to be 'No, it will not fail' and 'Yes, it can be built'. These categorical answers may be difficult to give in practice. There often needs to be some analysis to develop a level of confidence with which to make the choice. The amount of analysis required determines the level of understanding and experience needed to provide the evaluation of the suggested solution before the decision is taken to put the final proposal into practice.

It is this basic ability to suggest how a building may be constructed and then to carry out an evaluation asking the questions whether it will fail and whether it can be built that is developed in this book. The evaluation points to changes in the suggestion, which after re-evaluation will lead, through a series of refinements, to the specification and details to be adopted. This requires knowledge of what potential solutions may look like, with current practice and precedent as the major sources for an initial suggestion. It requires an understanding of what is necessary to be specified to describe the proposed construction in sufficient detail to carry out the evaluation. The ability to carry out the evaluation, on the other hand, is dependent on an ability to 'see' the proposal working in the dynamic systems of the physical and social conditions in which the building is to be built. This process of choice and the way it leads to the identification of the knowledge required is shown in Figure 1.1.

It will be shown later that the process of choice is not just an analysis of the behaviour of physical performance. This evaluation of a proposal's response to the dynamic physical, chemical and biological systems of nature is vital, but building also takes place in a social, legislative and cultural context. There will be an imperative to ensure that the resources and know-how are available to manufacture and assemble the building safely, economically and

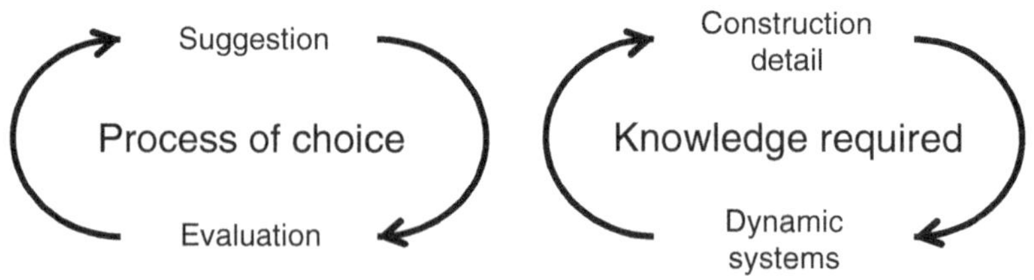

Figure 1.1 Relationship of process of choice to knowledge required.

in a timely fashion. This will require knowledge of the available industrial systems associated with on-site and offsite production, in addition to the availability of labour and craft skills. Further, it will be necessary to ensure that the cost of the solution is monitored, and that its social and environmental impact is audited. These will demand an understanding of both the economic and social systems in which the building is to be created and operated.

1.3 The initial suggestion

If the process is suggestion and evaluation, and the starting point is suggestion, it is necessary to know how we make the initial suggestion. How do we make that first best guess so that we can get started? In most cases the suggestion is informed by precedents. A precedent is a building or building detail that serves as an example of 'good' design and construction. Something, somewhere, has been done before that gives clues as to how the new solutions may be formulated. The precedent provides knowledge of solutions and how they perform in practice, i.e. how the building weathers over time and how the building users interact with the building. This knowledge can be used to inform or suggest subsequent actions, feeding into design possibilities and probabilities.

In times where change is limited the circumstances will have been faced several times in the past and successful solutions will have evolved. The well-tried and tested solution needs only to be suggested and, with some evaluation to ensure the circumstances have not materially changed, can be immediately adopted. If the circumstances are changing, current solutions may still be the best starting point. They represent not only a sound basis in performance but also an existing base of resources and know-how to manufacture and produce the materials and details. Evaluation may modify but not fundamentally change the solution, which is best described as incremental innovation. This, over time, gives rise to several general forms, known as typologies, from which specific solutions can be derived. A building language, a vernacular, has been developed that forms the precedent on which subsequent, informed, decisions are taken. Building design precedents can be found in the architectural literature that focus on a particular building typology, such as a school, a hospital or housing. Building technology precedents can be found in the building technology literature, such as the companion series to this book: *Barry's Introduction to Construction of Buildings* and *Barry's Advanced Construction of Buildings*. These books provide examples of how buildings could be assembled, not how they should be assembled. They offer tried and tested solutions that provide the initial suggestion to be evaluated and improved.

In some cases, perhaps where there are increasing user demands, or the structure of industrial practice is changing, then similar or related practice may have to be investigated to test and develop new solutions, for example, looking to other industries to see how technological change is impacting manufacturing. On the occasions where suggestions must be derived from little or no existing previous work, a more fundamental understanding of the behaviour of the construction will have to be applied as a way of mitigating risk. This is exactly what our response to the climate emergency requires: a fundamental rethink about the precedents that we have relied on and a return to the underlying principles of design and construction technology. This suggest a mixture of radical innovation through technological development (hi-tech) and a return to some of the earlier principles of design and construction that were predicated on limited resources and the need to recycle (low-tech). Fundamental decisions, such as to reuse an existing building rather than to demolish it, can have a major impact on embodied energy and material use. As an approximate guide, repurposing buildings has a 50% lower embodied energy use compared to demolition and new build.

While experts approach making suggestions based on their knowledge and experience, it is, in many ways, not necessary to know much at all to make suggestions and get started. It is the evaluation that requires the expertise. Novices or casual observers may make suggestions that may be hailed as brilliant observation, but in truth they have no way of knowing whether that suggestion is workable or not, or whether it will stand the test of time. It will take the expert to spot its potential and prove its worth through evaluation. The power of the expert to spot potential is probably associated with the ability to carry out a rapid, approximate evaluation before subjecting the suggestion to more rigorous and explicit analysis, using digital tools to calculate, model and simulate the outcome. This provides an estimate of the expected performance, which, in an ideal world will be very close to the actual performance.

1.4 Carrying out the evaluation

The heart of the success of the process of technological choice lies in the ability to be able to carry out an evaluation. The need to carry out a series of analytical exercises determines the knowledge required and manner of its application. It requires a level of understanding to answer the question 'What if we build the building as proposed?' and will involve several professionals and specialists to provide the answer(s).

While the suggestion and the ultimate solution describe the construction in what appears to be a static detail, the evaluation of the suggested construction must describe a dynamic system of behaviour (Will it fail? Will it perform?) and a production process (Can it be built? Can it be disassembled?). The process starts with the client's brief, which will inform the development of the design by establishing performance criteria. The design will be developed as a response to the social and physical context in which the building is to be built. It is against these criteria that performance will be judged. There will be many possible ways to construct a building to achieve the performance levels expressed in the client's brief. The criteria for choice of the technical solution comes from the identification of the function of the parts and how they contribute to the function of the building as one entity. The whole should be more than the sum of the parts.

It is too easy to see technology as only being the final construction, an assembly of components and materials that forms the building as a static object. However, making the choice of what construction should be adopted has to be rooted in an understanding of the building as a dynamic system responding to changes in conditions and open to a failure to perform as intended. It is necessary to be able to envisage not only the physical construction but also the building's behaviour under the conditions it will have to endure during its service life. Both are of equal importance. If the suggestion is not visualised correctly, its behaviour under analysis may be misinterpreted. Some specification of performance must be established and then the suggestion must be tested in the mind to assess the risk of failure within agreed design conditions. If the dynamic system of behaviour is not visualised correctly, a flawed proposal may be adopted. Picturing the building as a dynamic system will involve identifying the flows and transfers that take place both within the building and through the construction fabric, as explored further in the chapters that follow.

Technological choice demands skills in these areas of visualising the object and the systems, and then the conceptual manipulation of the systems acting on the construction to predict behaviour and assess the risk of failure. If the two basic questions to ask of a proposed solution are whether it will fail and whether it can be built, the criteria for choice come from an understanding not only of the potential dynamic flows and transfers when the building is in operation but also of the resources available. Performance can only be realised if the resources are available to construct the building. This relies on the manufacturing and assembly possibilities but will also include the existence of design expertise and the options for maintenance and recycling. Knowledge of available techniques and know-how for production is crucial to the choice of the final construction if it is to be successful as a building design and as a completed building project.

As a design concept emerges, it is necessary to question what construction solutions may be used to fulfil the design requirements. It is then necessary to question whether this solution is available with current technology and resources within any environmental, cost, or time parameters. The resources available make the design a reality. Technology stands between the design of the building and the management of the resources. Choices can be made that may extend current production and design knowledge, but this must be recognised, and any costs involved in prototypes or training and the risks involved must be accepted before the final choice is made. The design and resources available condition the choice. These two areas are shown in Figure 1.2. These are the two areas that must be understood before any analysis can be started leading to the final choice of construction.

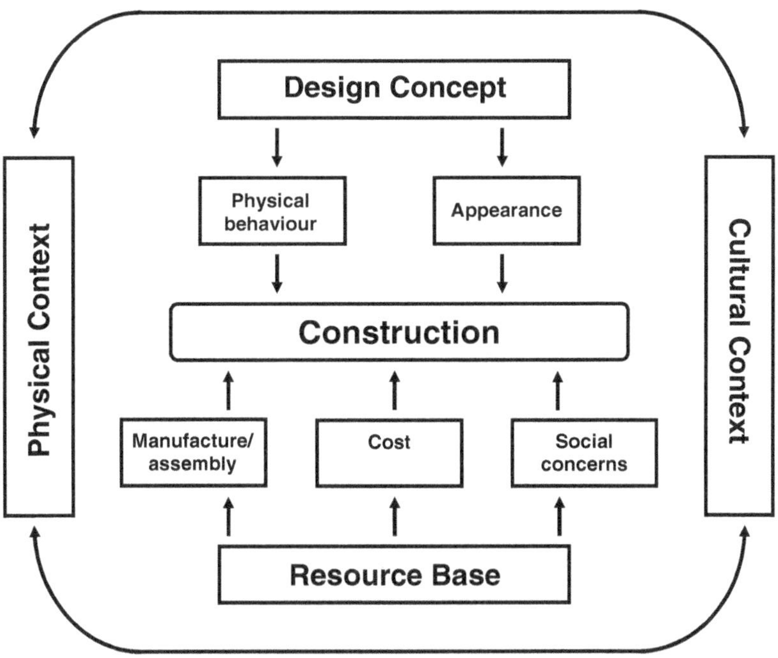

Figure 1.2 Framework of analysis.

1.5 Physical and social context

Figure 1.2 also indicates that before any choice for a specific building can be made it is necessary to understand something of the context in which it will be constructed. There needs to be some knowledge of the conditions that exist when and where the building is to be built, possibly with some assessment of how conditions may change in the future.

Construction takes place within, and has an impact on, the world in which it is undertaken. Some description of this world is necessary for both technical and socio-economic analyses. The world can be represented as a series of contexts, both physical and cultural. The physical context includes nature and climate. These exert forces that act on the building; they provide the raw resources and may be adversely affected by the construction and operation of buildings. The physical context includes the surrounding development and the need for the design to respond to existing buildings and spaces. People and their social, economic and political systems create the cultural context. It needs a response to local, national and global needs, based on beliefs and fundamental world views of the relationship between individuals and between society and the other components of the natural world.

The interaction between these two sets of contexts has been brought into focus by the need for sustainable development. The realisation that development cannot continue without consideration of its impact on the natural environment as well as established economic and social considerations calls for new knowledge associated with materials choice, energy use and waste disposal aspects of our chosen construction. There are, therefore, dangers in seeing these two sets of contexts as separate. However, it is still useful to consider them as having different dynamics. To resolve technical questions it is necessary to understand the physical context; and to evaluate the chance of a solution being successfully applied requires an understanding of society, its economic and political systems.

1.6 The basis of analysis

Having identified the design concept and the resource base as providing the criteria for choice, and having recognised the need to understand the physical and social context in which we build, it is possible to identify five areas of analysis (shown in Figure 1.2). The design concept is the translation of the physical and social needs for the operation of the whole building into a scheme that identifies the function and performance of each of the parts. To achieve this, the design concept must articulate both the arrangement and appearance of the spaces and the technical contribution each part of the construction will make to the creation and maintenance of the internal conditions. Two tests indicated in Figure 1.2 must be applied to a suggestion to see if it complies with the design concept:

- ❑ Does the physical behaviour of the construction provide a building fulfilling its functions to the required performance level?
- ❑ Does it provide the right attributes of appearance?

The test for physical behaviour involves three separate areas of analysis:

- Creating environments
- Under load
- Over time

Three tests indicated in Figure 1.2 should be applied to a suggestion to see if it is achievable with available resources:

- Can it be produced, including its manufacture and assembly and the subsequent processes of maintenance and recycling, with existing skills and know-how in a reasonable time and at the required quality?
- Are the resources available at reasonable cost?
- Will it be compatible with the prevailing environmental and social concerns?

These resources are both natural (from the environment) and social, so both aspects must be part of the evaluation. It is only possible to use resources that nature has provided. However, the level of economic development in a society will provide the capacity to process materials, develop the skills and machinery to work them, the intellectual capacity to undertake design, and the provision of capital to invest in the enterprise of the construction. Both the technical properties of materials and the existence of the knowledge to exploit them, plus any environmental impact of their use, must be analysed in evaluating a suggested form of construction.

These areas of analysis to be undertaken are, by and large, unrelated to each other and each needs its own tools of analysis and knowledge base to be successfully applied. Carrying out one analysis may indicate the need for a change in the suggestion, but the change that is chosen may make the analysis of a previous aspect invalid. It is not until all the relevant areas of analysis are shown to be satisfactory that the suggestions can be adopted as the solution. The analysis must be undertaken for all aspects of the construction, from the overall structural system to the finest detail such as the screw that fixes the final fitting to a wall. The work in this is clearly overwhelming if it must be carried out for every building that is constructed. It follows that designers take shortcuts and rely on solutions that are known to work, and which informs much of the process of choice.

One of the most important decisions that experts take, of all the thousands of choices that must be made to fully specify the construction, is which of the areas of analysis for which parts of the building are significant. Where, if analysis is not thought through, is the greatest risk of failure? It is not clear how experts make this judgement. One possible explanation is that while looking at a suggestion they make many rapid checks against the seven criteria. From their knowledge and experience they take a view on what is well within failure limits and where the risks of failure are higher.

1.7 Knowledge needed for choice

It is now possible to start to identify the knowledge that is required and the areas of understanding that must be developed to carry out a full evaluation that considers the entire life of the building in a circular construction economy. Although used in various combinations in the different areas of analysis, it is possible to put forward a tentative list of types of knowledge that will be required:

- Setting the performance levels as the criteria against which the evaluation will be made.
- Defining the conditions under which performance must be achieved.
- Determining the fundamental behaviour of the construction that would lead to failure to meet performance requirements.
- Identifying the materials' properties that will govern the behaviour and performance of the building as one entity.
- Thinking through the behaviour of the combination of materials and details under the conditions envisaged.
- Identifying and simulating the process of manufacture and assembly, together with the resources required to produce the building within quality, time, cost and safety parameters.
- Simulating the process of disassembly and materials recovery at a specific, if uncertain, point in the future.
- The cost in economic, social and environmental terms of using a particular form of construction to meet performance criteria.

The process of evaluation requires the building to be conceptualised in several ways. While the final building will be visualised as a physical construction, it is necessary to perceive the building in several other ways when carrying out the evaluation. Initially, the construction must be seen as fulfilling a set of functions with associated performance levels. For evaluation, the building should be viewed as a series of physical systems responding dynamically to changing conditions. The building then must be seen as a series of production operations with the resources required for the realisation of the design. All these will have economic, environmental and social implications that must be understood as part of the decision-making process. When all of these can be mastered, choices can be made associated with an assessment of risk and reward to realise a successful building.

In the chapters that follow the intention is to provide a logical sequence of events to enable analysis and choice. Although the chapters are designed to flow from one to another, they are also designed to be dipped into, and readers may well find themselves going from one chapter back to another as they explore various themes; much like one would do in practice when reviewing information and making a choice following a series of iterations.

Chapters 2, 3 and 4 are designed to set the scene. Chapter 2 provides a narrative surrounding the development of performance criteria. The design concept is linked to making choices about the construction, which is explored in Chapter 3. This involves the consideration of construction variables, which is the theme of Chapter 4. From this base it is then possible to consider how physical and social conditions are defined (Chapter 5), the resource base that influences choice (Chapter 6) and sustainability in Chapter 7.

Chapters 8, 9 and 10 are dedicated to different aspects of behaviour. Chapter 8 explores the inter-related aspects of the environments that we create. Chapter 9 focuses on the physical behaviour of the building under load. Chapter 10 looks at the building's physical behaviour over time. Collectively these chapters cover the main phases of the building's life cycle. Once this is understood, it is then possible to think about interface design (Chapter 11) and the processes necessary to assemble the building (Chapter 12).

Part 2 of the book explores choice and challenges readers to question standard construction solutions. Chapter 13 describes how we could use the framework to question how we currently build. This is followed by chapters that aim to confront embedded behaviours and established construction solutions. The subjects tackled are foundations (Chapter 14), cavity walls (Chapter 15) and windows and doors (Chapter 16). These three chapters conclude with reflective exercises to be tackled by readers by themselves or as a group exercise in the physical or virtual classroom.

Building Purpose and Performance: Chapter 2 AT A GLANCE

What? The purpose or function of the building is to support the activities to be carried out within it, using both physical and social performance criteria. Performance criteria have evolved as buildings have become more sophisticated and users have become more demanding of their buildings. This has helped to emphasise the importance of performance and has also resulted in a better awareness of the difference in anticipated performance and actual performance; known as the performance gap.

Why? Choice of construction can only be made in the knowledge of the environments in which the construction operates and an understanding of the mechanisms or actions that lead to under-performance or failure. Both physical and social functions of the building lead to physical performance requirements covering environmental, structural and reliability aspects of the construction. These requirements lead to performance expectations and hence definitions of success or failure. The other factors to be considered in making the choice of construction include production resources and know-how, cost, legislation and the need to consider the whole life of the building and its environmental impact.

When? The purpose and performance of a building are best established at the briefing stage, and subsequently tested, adjusted and reaffirmed as the design evolves from early ideas and concepts through to the confirmation of materials and components in drawings and specifications. The building will be assembled based on the information provided by the design team. When the building is complete and in use there will be an opportunity to assess its performance against it predicted performance. This will be achieved by conducting post-occupancy evaluation, using the performance criteria as the benchmark to gauge performance in use.

How? Performance criteria are established at the early stages of the project and are a manifestation of the client's wants and needs. They represent the values of the society in which the building is to be used and must conform with prevailing legislation and codes. Performance criteria are usually set out in the project brief, and they serve to inform the development of the design, its detailing and its construction.

2 Building Purpose and Performance

Technology is used to mediate between the world as it is and the world as we want it. This chapter expands this idea, focusing on the users and their activities, introducing the functions expected of the building and how these must be translated into a process of choice for the construction of each part. The focus is on the development of performance requirements. Performance identifies the level to which the construction must function under specific design conditions and becomes the criterion for any evaluation of success. Visualising the conditions under which the construction must function, and the expected response of the construction chosen provides the basis for an analysis to evaluate satisfactory performance. This chapter explores the development of performance criteria and its relationship to a more sustainable and resilient built environment. The challenge of the performance gap between intent and reality is discussed before turning to analysis of physical behaviour.

2.1 Activities, space and construction

Buildings are characterised by their overall purpose or function, be it a house, factory, hospital, or office. Because of this they are often immediately recognisable, although their outward appearance and their interior layout may vary. To understand the form of the building, it is necessary to understand the day-to-day activities that make up the function for which the building is procured. The design, engineering and subsequent construction (or production) of the building will have to play its part in ensuring the wellbeing of these activities. The notion of wellbeing is normally associated with the people who use the spaces, but the building will have to protect the equipment being used and the storage of items associated with the activities. Wellbeing of people also needs to include the requirements of the owner and the wider public, as buildings are social institutions and create the (social) spaces that surround them.

Most spaces within buildings are designed to accommodate people, and consideration of the users' wellbeing becomes a dominant aspect of the performance of the construction. In some cases, such as the rooms for the early computers and specialist storage areas such as vaults and cold stores, the equipment housed in the building, or the storage conditions dominate the requirements. Designing for the wellbeing of people is essentially concerned with health, safety and comfort. Health should be understood as meaning both physical and mental health, both being a function of physical and social conditions created by the building. The owner may see a purpose beyond the simple support of activities, to include image

Barry's Construction Technology: Analysis and Choice, Third Edition. Stephen Emmitt and Tony Bryan.

and status. The surroundings in which the activities are undertaken have social significance and must form a feature of the design.

Each society over time evolves common building forms for each purpose type that become recognisable within that society because of their size, arrangements of internal spaces and social statement. With this emergence of building form there is a need to develop appropriate construction solutions. This process is dealt with in more detail in Chapter 3, but first this chapter explores the complex range of performance issues. These will be established predominantly by the users' activities, although later by the owner and then by government in the form of legislation on behalf of society.

2.2 Introduction to performance requirements

To explore the range of performance issues it is useful to engage in a mind experiment. The situation to be imagined is one where civilisation as we know it does not exist and the task is to provide shelter. The scenario is one of few resources of materials, labour or knowledge. There is also little development of an economic system, so the whole design and construction process must be undertaken by the users. This is an early example of self-build. The mind experiment starts by asking which performance the user of the building would seek to satisfy first. The experiment can then be extended through time, assuming increasing resources and socialisation, introducing additional requirements that would add value and utility and improve user comfort and wellbeing. By undertaking this mind experiment it will be possible to introduce the needs of the user and hence the functions that individuals will require the building to fulfil. As the user is also, at least initially, the designer and builder of the shelter, they must decide on its construction. The experiment will, therefore, also start to introduce the ideas and knowledge that will be needed to make decisions on the construction of the building. All the issues and factors introduced here are relevant today.

In the beginning

Given no immediate risk from attack, we can speculate that people would organise their technological endeavour in the following way. Thought would have to be given to the range of activities of everyday life that should be housed in the shelter. Initially this is likely to be very few, perhaps only sleeping and cooking, with most activity being undertaken outside depending on the prevailing climate. This will provide a definition of size and any division of the internal space that needs to be made. It would be immediately recognised that these requirements for size and shape would be limited by the resources available locally. Decisions would have to be made as to how much effort should go into developing the construction to match aspirations and satisfy immediate needs. The priority in terms of establishing functions for the building fabric would be to remain dry, a condition fundamental to the health and comfort of people, as well as the storage of many natural products. This would be accompanied by the requirement to be warm, but it is possible that this will initially be achieved in combination with other technologies. These would include clothing and the use of fire, which at this stage may or may not be associated with the construction of the shelter.

As soon as any construction is contemplated the 'builder' is immediately faced with achieving a stable structure and making some judgement on how long it may last. Although buildings may be erected for environmental control, their very existence demands considerations of stability and reliability if they are to be deemed satisfactory by the user. This will require the identification of possible modes of failure. At this early stage this would include how the construction might leak or collapse, much of which would be based on trial and error. This understanding of the ways in which performance may be lost becomes crucial to directing effort to significant aspects of the solution. This would also lead to some questioning of what maintenance may be necessary and what parts may have to be renewed to continue the performance and functionality of the building until such a time that the artefact was no longer serviceable and had to be abandoned or demolished to make way for a new building. To minimise unnecessary effort, the materials that were still serviceable, such as stone and timber beams and posts would have been reused in a new construction.

It would also become necessary to have some knowledge of the environmental conditions in which the construction has to maintain performance. This local knowledge would be required to consider not only the environmental control but also the stability and reliability of the building. These three criteria become the key physical performance issues. External climatic conditions such as frequency and volume of rainfall will need to be assessed for the weatherproofing function. Damaging wind speeds and prevailing wind direction would need to be assessed for stability and ventilation of the interior. It would, however, soon become clear that wind is a major factor in weatherproofing. Some of the most arduous weatherproofing conditions come with wind-driven rain. It is not possible to say what aspect of the environment needs to be known to assess durability until some choice of the basic materials is made. Specific decay agents in the environment will degrade different materials at different rates. Linking materials to their generators of decay is just one more question that must be answered in the quest for knowledge to make sensible choices for the construction. This assessment of external climate would soon reveal that often the more extreme conditions of open land are ameliorated by local natural features, that exposure can be modified by the position and orientation of the building and its openings, which will limit the demands on the construction and/or give better internal conditions.

The next steps

As the number of activities carried out in the building increases, the provision of good air quality may be the next priority for improvement of the construction. This is probably the next requirement for health and comfort after dryness and warmth. Seeking a solution that would improve the quality of the air would involve an understanding of ventilation since there is an abundance of good air outside. It would have to be recognised that the air outside may also be cold and that this will make the requirement for warmth more difficult to achieve. This may mean that a solution for good air may only be possible when air temperatures outside are sufficiently high or air quality inside is particularly poor. Assuming that the active technologies of building services are not available, modification would be necessary in the existing construction. It would be necessary to use natural ventilation driven by wind pressures around the building. Openings would have to be introduced to ensure the exchange of air. Their position and size would have to be considered. It would be possible to introduce a level of control by providing shutters. The openings could eventually be as sophisticated as chimneys, but history tells us that these often came much later.

It is important to note that the containment of the bad air is a direct result of employing construction to remain dry and warm. The air outside the building is almost certainly good. This realisation that the construction can create undesirable conditions should not go unnoticed. Changes introduced to fulfil new performance demands may change conditions that have always been satisfactory. This happened towards the end of the twentieth century when a requirement for energy saving was satisfied by the introduction of insulation. Under some environmental conditions this led to condensation that has a deleterious effect on the basic requirement of keeping dry. This led to health problems for the users of the buildings.

The introduction of new performance requirements may also introduce failures in the construction that have previously not occurred. This happened in the condensation example, where dampness caused not only health problems but also a deterioration of the construction materials themselves. In this mind experiment it is possible to speculate on a similar problem with the improvement in internal air quality. Given the lack of social organisation to produce processed materials the construction would almost certainly be made of relatively unmodified organic materials. In this case it may be speculated that if the air contained a large amount of smoke from cooking and heating, it would have permeated the construction. This would have effectively fumigated the organic materials and kept infestations or decay-promoting fungi and insects at bay. With the development of technologies to improve indoor air quality, it would have changed the conditions within the building fabric, which could allow some of the natural agents of decay to thrive, thereby reducing the life of the construction.

More physical requirements

This mind experiment continues based on the assumption that no major threat to security arises. This would inevitably lead to a diversion of resources and may change the priorities of individuals or society in seeking improvements in their buildings and it may also lead to new types of buildings. Given peaceful social development the next priority may be difficult to identify. It may be that the infestation of insects and small animals that will be attracted to the improved internal environment becomes a problem. It may be that in the harsh realities of life, and in the absence of any realisation of the health risks, these infestations would be tolerated. Measures to eliminate the larger animals, if not already considered as a danger, may now become important. At some time, however, infestation control would have to be included in the list of functions required of the construction to improve wellbeing. This is part of the more general requirement to keep the enclosed space clean. Internal surfaces that are dust-free, smooth, and wipeable become desirable attributes where cleanliness becomes important to the health and comfort of the user.

Having gained good clear air, more activities could be carried on indoors. It is, therefore, possible to imagine an increasing need for light to improve comfort levels and further extend the range of activities in the building. The choice is between natural and artificial means of lighting. Natural would involve openings, while artificial would only demand a safe place to support what would almost certainly be an independent light source. Although independent of the building, it would have an influence on it. Perhaps the major influence it would have relates to an increased risk from fire. While it may not be possible to improve the fire resistance of the whole building at this time, it would be sensible to employ non-combustible materials at least where the light source is to be mounted. Here we see the increased use of technologies independent of the building, seen by the user as an asset to wellbeing and influencing the required functions of the building.

Another requirement is for some measure of acoustic control. This would, of course, only become a concern if the external level of sound were somehow interfering with the overall sense of wellbeing, leading to the perception of unwanted noise. It may become a problem when the building has separate rooms for specific activities. Noise from one activity may well be incompatible with the conditions necessary for another. Noise is a particular problem of civilisation, and especially for those societies that have advanced technologies and where there are high levels of people in one area, such as cities. It also seems that it has a cultural basis in attitudes towards individual rights, privacy and intrusion.

The technical considerations become more complex

It is useful to view the need to consider noise performance from a different perspective. It may be that there is a need to modify the sound between outside and inside, yet the construction used for the other requirements already has sufficient, if not excess, ability to perform as a sound barrier. With low performance expectations the final form of construction is dictated by only a few of the functional requirements. As more functions are identified and the level of the required performance increases it becomes difficult to provide simple construction solutions. Composite solutions are necessary with different materials to satisfy different functions at the levels of performance that are required.

The analysis of exactly how each part of the construction is interacting in providing the overall function becomes more difficult. The danger comes when changes are suggested. It is necessary to ensure that the change does not make a previously satisfactory performance critical so that the construction fails to perform, in an unexpected way. The trend towards lighter partitioning in housing has, by reducing the weight and changing natural vibration frequencies, reduced the sound transmission resistance. In the past this was normally more than satisfactory when partitions were thicker or made of denser material. The lighter, cheaper solution may become critical in some situations where the soundproofing performance of the heavier solution had always been taken for granted.

The need for composite construction leads to the search for materials with properties that match the specific functions. Processing and manufacturing of materials allows these properties to be more accurately achieved at scale. This not only changes the ability to gain performance but also influences the production resources required. Materials and manufacturing methods may also be changed to provide more economic buildings. The innovation may not be driven by performance but to seek production and cost gains. This may lead to performance failure in previously satisfactory solutions. Those responsible for the choice of construction must re-evaluate the possible performance failures if construction is changed for production or cost advantages.

Taking storage and equipment into account

The environmental requirements that have been introduced are those necessary to maintain the wellbeing of the people who use the building. These can be summarised as: dryness, warmth, cleanliness, light and quietness.

Observations of this list will identify that, with only slight modifications (e.g. warmth needs to be more generally interpreted as temperature control), it could equally well be applied to maintaining the wellbeing of the equipment and stored items. The required internal conditions for storage and equipment may be different, even hostile to life as in the case of cold storage. They will still demand the same type of functions of the construction. There would, however, be changes in performance level due to the change in specification for the internal environment, even though the external condition may be the same. In the example of cold storage, levels of insulation would be different.

More significantly, the equipment and stored items, as well as the living occupants, will contribute to the internal environment with noise, smell, heat, fumes, etc. It may be that the construction is called on to moderate these effects by isolating storage from living areas by partitions. Independent technologies in the same room such as cooling of food may be applied using ideas such as evaporating water from a porous pot or a refrigerator. However, independent technologies may influence internal conditions. In the example of cooling food the porous pot releases vapour and the refrigerator releases heat. The dynamics of these changes in the internal environment due to occupants and independent technologies in the building must become part of the analysis. Sealing a building to reduce heat loss through air changes can lead to a build-up of vapour (if only from people breathing), which makes condensation on surfaces and within the building fabric more likely. One solution is to install and air-handling unit that can recover heat from the air as the air is circulated. This helps to maintain the temperature within the building while also ensuring adequate air changes.

Although the list of environmental requirements generated for people will cover most circumstances associated with the equipment and storage in the building, it may in some situations be incomplete. Some modern equipment has environmental operating requirements, such as the need for anti-static precautions, that may not currently be considered necessary for the wellbeing of people. What must be achieved is a balanced solution to ensure the wellbeing of both the participants and the processes associated with the activities for which the building is required.

What should be considered tolerable?

Setting performance levels involves considering what is tolerable. And this results in some fundamental questions. What should be considered as normal limits to be maintained in the internal conditions? What external conditions should be allowed for in the design? These questions must be answered in terms of risk and consequences. The risks will vary from those that may occur very rarely and those that are more common. Extremes that happen rarely can have serious consequences for wellbeing and the activity that takes place within buildings, sometimes threatening life and property. The more frequent events, even with lesser consequences, also need to be considered. This is not independent of the economic resources available and prevailing social attitudes. In many parts of the world, earthquakes will occur, but the construction to minimise the risk to life may be seen as increasing costs beyond that to be extended to all sections of society. Similar circumstances have existed in the past with fire safety and building. Increasing wealth, effectively distributed in society, will affect what is considered tolerable and will be reflected in more demanding performance expectations and building codes.

Generally, wide limits would be accepted for solutions at the beginning of our mind experiment, and as societies evolve and develop, they will seek to extend the range and level of demands. This often includes a requirement for a greater level of control as it defines a narrower band within which performance is deemed satisfactory. These trends place more demands on the solutions and failure is more likely.

Some extremes that happen rarely need to be considered if they have serious consequences. One risk that comes from the internal activity that generates extreme conditions has already been raised. It is the issue of fire within the building. At the beginning of the mind experiment, while escape for people would be required, it is unlikely that there would be any expectation on the part of the user that any of the building would remain after a fire. Although fire is now rare in modern buildings, its consequences are such, particularly in risk to life, that it is currently deemed necessary for the construction to continue to maintain a selected range of performances both during and after the fire. This would almost certainly include stability, although insurance requirements may extend the performance range to other factors.

External climatic extremes such as wind, flood, lightning and even earthquake may be considered via performance criteria. Design for flood prevention will be required for buildings sited in areas prone to flooding, for lightning mitigation for tall buildings, and for earthquake if the building is in a seismically unstable region. New threats created by increasing sophistication of solutions are now emerging. Draught-free homes created to conserve heat may suffer from a build-up of radon gas if built in an area of the country where it occurs naturally in the rocks below the building. The chemical contaminants in the soil on sites previously used for redundant industries threaten both the occupants and the fabric of the building. In both examples measures will need to be put in place to mitigate impact.

Introducing social requirements

Having established the requirements for health and physical comfort, it is now necessary to consider the social aspects. The consideration of performance for more social aspects does not come only when physical requirements have been fully satisfied. Human endeavour very quickly becomes a social activity and there grows a need for social organisation, establishing cultural norms, provision for leadership and political activities and an economic system for the exchange of goods and services. Buildings play their part in this process not only to house the activities but also in that they reflect, affirm and even become instruments for change as the cultural and economic order develops.

As well as the issues of status and image, to be discussed later in this chapter, two other functions associated with wellbeing arise when society starts to develop. These are the requirements for security and privacy. They are mainly associated with mental health, although the security risk, if it becomes a reality, may involve physical danger. The point at which people would take these functions into account would be dependent on the nature of the society that develops. They are not as predictable as the physical needs and may well have to be addressed before the desired level of physical comfort and safety is achieved.

Security requirements will be dependent on the existence of a perceived threat. If the risks are seen as sufficient, technological ingenuity, time and resources will be devoted to combating the threat if social devices such as the rule of law or treaties cannot be achieved. Threats come from both inside and outside a society, each of which poses different challenges to the technological skills of the builder. In the extreme, in buildings such as bunkers, the security threat dominates the design as the technology of survival becomes pitted against the technology of destruction.

The requirement of privacy is much more rooted in the cultural attitudes to community. It is cultural interpretations of what activities should be shared and those that should be carried out in private that dictate the need to introduce this performance into buildings. Performance levels of privacy can be identified associated with the senses, visual and aural being the two main requirements of the building influencing both the external envelope and normally demanding some division of internal space.

The functions of the building that have a social origin often translate into aspects of physical performance already introduced. Security makes structural demands on the fabric. Privacy determines many of the sound transmission requirements of construction.

Influence of production resources

As more functions are included and more materials are introduced, production resources and know-how must develop. The relationship between the need for a building and the availability of the materials, skill and tools to produce it is always a determining factor in the choice of the construction. Access to production knowledge and expertise limits the pace of development of new solutions to achieve new performance. However, developments in materials and production can also suggest different ways of achieving similar performance.

This mind experiment has taken the perspective of the purpose of the building and requirements of the user. It is easy to forget that the means of realising that performance may need as much thought and ingenuity as the design and detailing of the fabric and structure of the building. In many old buildings it is easy to see how they work but often difficult to see how they might have been built without appreciating the tools and skills available at the time. In many innovative buildings, detailing is often determined as much by manufacturing or assembly procedures as it is by satisfying performance. The availability of manufacturing and assembly resources also determines the cost. The need to develop production procedure affects the time required to construct the building and determines the risk associated with a particular solution. All these factors contribute to the final quality of the building and its success judged by performance in use.

Considering cost and value

Throughout this mind experiment the question of cost and value will have to be considered. Buildings are usually costed in terms of money, but there are also considerations of cost to the environment and on the impact that they may have on society. The resources available to any society are limited and decisions must be made as to how these should be allocated. As part of the culture these are the political decisions involving both the vision of the type of society that should emerge and the preferred methods of achieving it. It is from this cultural process that demand is initiated, the cost of resources determined and value assigned.

Buildings will only be commissioned when the cost of a building that performs to an adequate standard is considered good value. If the costs are too great, decisions must be taken on eliminating functions, reducing performance or changing production methods. These changes should ensure that the building is still considered as value for money. Assigning this cost requires that at least a broad outline of both the construction and the production methods is known. At some point in the process the decision to proceed to full design and production based on estimated cost must be made. For buildings that can be produced using well-developed current forms of construction this decision can be made very early with some certainty. For a project where there are few precedents, possible solutions may have to be investigated before cost can be established with sufficient certainty to take the decision to proceed. It is, however, always a requirement to set cost limits as early as possible and then develop the final solution with these costs in mind.

Building for status and image

Status and image will at some time in the development of society start to influence the requirements of the building. This may be reflected in the size of the space provided over and above that strictly required for the physical satisfaction of the activity. It may be associated with the quality of the materials used being acknowledged as expensive. These expensive materials must have the ability to perform their physical tasks without failure, but they could have been replaced by a less expensive alternative. These requirements are often overlaid with issues of art and fashion and symbolism. The influence of these should not be underestimated, but a full discussion of their nature and form is inappropriate in this text.

The pressures of status and image are not just related to statements of wealth and position. We all need to make statements and feel comfortable about our place in society related to the prevailing cultural norms. It must be recognised that these cultural norms can be highly regionalised, being associated with tribes or nations, although improved communication and movement of people seems to reduce this variability. We all care about what our buildings look like and the environment they create. Using and being associated with buildings that match the psychological demands of the activities carried out in the building is very important to our mental wellbeing.

These issues become associated with aesthetics, the visual and deeper, even spiritual, quality of beauty of both natural and manufactured articles. These are the aspects of the world that satisfy our feelings rather than our physical needs. Buildings intrude into the rural landscape and create the cityscape. Their shape, form, colour and ornamentation have a significant impact on our lives.

Control from society and legislation

All social groups have cultural norms of behaviour that limit the actions of individuals and impose sanctions for transgressions. This eventually becomes formalised into legislation, some of which will start to impinge on the choices that may be made for construction. Direct legislation is normally associated with public health, limiting the use of scarce resources or the position and appearance of whole buildings. Once manufacture and assembly start, there are laws associated with the health and safety of operatives and the people in the surrounding area. All of these will have to be acknowledged and satisfied in the final choice. If the choice of construction is made in the spirit of responsibility within the cultural norms, it should not dominate the solutions chosen. Legislation is for the protection of those who use the buildings and should, therefore, be easy to accommodate into a process whose objectives include satisfying social and individual needs. More recently legislation and guidance has expanded to include our impact on the planet.

Laws and legislation can be slow to change and may be drafted such that its interpretation becomes a matter of opinion and interpretation. These factors can spur or inhibit technological change and those responsible for the choice of construction must be aware of the law and its interpretation. There is much legislation that will indirectly influence the choice of construction, although predicting and understanding the links are more difficult. Contracts become necessary between the organisations that undertake different parts of the building process. This legislation affects decisions on the construction and what risks they carry. The knowledge and experience that the organisation has then influences the choices made for the construction of the building.

Clients, users and change of purpose

Another complication that arises as society develops is that individuals who become the owner/occupiers commission fewer buildings. It is now possible for those who commission the work to be an investment company which passes on the task of operating the building to another company for use by yet another group of people. This does not nullify the idea that satisfying the needs of the user should govern the performance of the construction solutions. It is the provision of space that ensures that the desired activities can be carried out efficiently and effectively by the user that gives value and utility to a building. Those who commission the building may have objectives such as investment, but these will generally only succeed if the purpose of the building and its performance are satisfactory.

Even when relatively simple relationships between the client and initial occupation exist, the purpose of the building is likely to change with time. These can be quite dramatic changes when the building ceases to have a use or value to society. For example, the conversion of redundant industrial warehouses to retail and housing. However, most changes are much more subtle. For example, changing practices in medicine has rendered many hospital buildings inefficient and the digital revolution has changed the basic activities carried out in many offices. Increasingly the operators of buildings (as opposed to the users) see the building as part of the business's efficiency. This gives different value to maintenance, alterations and refurbishment, and makes the building a more integrated part of the facilities as one entity. This leads to the questions about the whole life costs of the building. More expensive initial options may be better value over the service life of the building, reducing operating costs and increasing efficiency and user wellbeing. This requires an understanding of life-cycle costing, which is usually linked to life-cycle analysis of materials and buildings.

2.3 The performance gap

With increased demands and requirements comes complexity and competing demands between a wide variety of performance requirements and the need to satisfy increasingly onerous laws and codes. The result of this complexity is that the completed building may not always perform as intended. This is known as the performance gap and is often associated with a difference in energy use between that modelled and that realised in occupation. The concept can be used more widely to cover other performance criteria, covering those that influence indoor environmental quality and building related issues, such as the way in which the building or elements of the building weather over time.

If the construction does not meet the requirements as set out in the building regulations, codes and standards, there will be a gap between the building performance and the regulatory requirements. In most countries this would need to be rectified by the constructor and the design team, however, it is not always evident until a problem occurs with the building at a later date, when the non-compliance is revealed. This could relate to the failure to install fire and smoke stops in the wall assemblage, incorrect installation of thermal insulation, etc. These are sometimes caused by negligence, but more often are a result of trying to meet many competing performance requirements, and not being able to satisfy them all.

The most common performance gap is related to the predicted energy use of a building and the actual energy use. This may be down to uncertainties in the modelling and or variations in how the building was constructed compared to the conceptual design. For example, a minor change in materials made during construction may have a significant impact on the overall performance of the building. It may also be down to how people use the building, which is not easy to predict and model. Individuals will have different levels of thermal comfort and may set the heating at a higher temperature than the building was designed to operate at. In such a situation there will be a difference between what was designed (intended) and what happens during occupation. There is a gap in performance and a gap in expectations. Trying to predict the behaviour of humans can be very challenging. Predicting the behaviour of the building is, by comparison, easier, however, the use of new materials and techniques in response to the climate emergency may increase the level of uncertainty in the modelling simply because there is not yet sufficient research available to inform simulations. It is inevitable that the performance gap will continue to exist and what designers and constructors need to achieve is as small a gap as possible given the resources available to match the expectations of the client, the building users and society. This is where digital technologies can make a positive impact, through real time monitoring of buildings and incorporation of learning into new design and construction projects.

To understand the reasons for the performance gap, the extent of the gap in performance, and hence the ability to address it in future projects, requires testing and monitoring. The testing of material is particularly important when using new materials or familiar materials in new situations. Some of this may be done in the laboratory and in specialist facilities, such as fire testing facilities, often supported by digital tools, such as finite analysis and through the use of computational fluid dynamics (CFD), to name but two common approaches.

2.4 Environmental impact as performance criteria

This long-term view of the building leads to wider questions about environmental impact and the notion of performance. Construction has always had a local impact on the environment. It exploits resources, changes ecosystems, consumes energy and creates waste. The society in both our mind experiment and now in reality has grown more prosperous; infrastructures have improved, allowing the growth of manufacturing and materials processing to serve an expanding population with ever higher performance expectations. With this has come a better understanding of the harm that humans are doing to their natural environment in trying to satisfy their needs and expectations through the act of building. The environmental impact of the technologies chosen for individual buildings will have local and global consequences. This has resulted in a call for a new way of thinking about the built environment, encompassed in circular thinking and regenerative design. The circular economy is based on a closed loop and efficient use of materials and energy, with little or no waste. Regenerative design goes further and seeks to revitalise and regenerate our built and natural world by taking a more considered and ethical approach to building. This demands a different lens to view the problem, and yet more analysis, with new knowledge required to make choices about the sustainable technology we employ for our buildings. It will lead to new solutions and materials being favoured as decision makers attempt to balance the need for innovation against the (perceived) risk of doing things differently. This is explored in Chapter 7.

2.5 Analysis of physical behaviour

With increased performance levels and demands comes the need to understand physical behaviour. Analysis of physical behaviour is predominantly a technical activity because it is undertaken after the function of the parts has been defined and after the conditions under which it has to operate have been established. It varies from an analysis of the appearance of the building in that it is not based on a social or cultural response but on a response to the condition of the physical environment. The analysis of physical behaviour must be based on an understanding of the laws of nature, the forces involved both in states of equilibrium and in the dynamic behaviour during changes in conditions. There are three primary areas of behaviour that are of significance to the success of the building, namely:

- Behaviour creating environments (Chapter 8)
- Behaviour under load (Chapter 9)
- Behaviour over time (Chapter 10)

This analysis of physical behaviour takes the idea of performance as its criterion for success. Every part of the construction fulfils a function and has a part to play in contributing to the function of the building as an entity. While the client or user determines the function of the whole building, the functions of the parts are determined by the chosen design concept. Having established a design and assigned functions to a part of the construction, it is necessary to identify a level of performance. It is against these levels of performance that analysis can be undertaken, and the failure modes identified. The language of performance levels will vary, but the common idea is that the performance statement must be testable. Examples would be that beams must carry load without exceeding a determined deflection, walls may have to resist the passage of heat with a specified level of insulation and no dampness should appear on any internal surfaces of the external elements of the building.

Preventing failure – a technical objective

Failure is a loss of performance. Failure follows when the response of the proposed construction does not meet performance levels. If the beam were too small or made from a material that were too elastic and had not been shaped or fixed to overcome these shortcomings, it might deflect too much. In such a situation, the beam has failed to perform. If the loads were larger than predicted and the beam deflects too much, the origins of this failure are in the prediction of the design condition.

Analysis of physical behaviour requires three areas of understanding:

(1) Identification of all the performance criteria
(2) Careful consideration of the source and nature of the operating conditions
(3) Understanding the mode of failure of the construction responding to working conditions

This requires an ability to imagine the ways in which the construction could fail to achieve performance under the design conditions. The clues to this should be in the performance specification, for example excessive deflection in beams under loading conditions or excessive heat transfer through the walls under winter environmental conditions. It will also require a method of predicting how much deflection or heat transfer will occur to check that behaviour will remain within performance limits. The clues to this will come from visualising and simulating the internal action of the construction under these conditions in order to determine the response. Future predictions of performance play a part in this analysis, and this is often difficult, given the expectation of longevity for most buildings. Establishing higher performance levels for some expected future requirements or conditions is problematic because the unknowns and uncertainties associated with higher performance levels will have cost implications.

These considerations for the future add to the difficulties of prediction. They may be included in the project brief but, if not, can be part of a technical analysis in so far as the sensitivity to error or change in conditions can be carried out. A solution that is sensitive to change may be less attractive if the future is not certain. This adds an additional dimension to the analysis. How close to failure is the solution when under working conditions? Will small changes greatly increase the probability of failure? This prediction of sensitivity can be part of the analysis undertaken by visualising the response of the construction under the expected conditions. Understanding of the response of the construction to the external conditions allows the estimate of behaviour under working conditions and any capacity it may have to withstand unexpected or deviant conditions.

2.6 The role of observation and science in the analysis

Experience can offer direct observation of construction both during production and under working conditions. In unfortunate cases experience can offer the opportunity to observe failure, for example by analysing a building collapse or a failure in the performance of, for example, a material or component. Fortunately, advances in software modelling have made it possible to simulate failure in a digital environment and test the proposed construction for its failure limits long before it is put into reality. Longer term performance under weathering is more challenging to simulate, and this still relies on accelerated testing in laboratories and field trials in addition to software modelling. If failure is observed, some explanations are necessary to determine what changes are required to achieve performance the next time. Requirements for higher performance, operating in different conditions, or a change in the construction for production for economic gain also require explanations.

For those engaged in technological activity the naturally enquiring mind is likely to seek such explanations. It is possible that over the past centuries, individuals have provided their own explanations to develop construction in a systematic way, refining solutions and occasionally making larger leaps of understanding to new forms of construction. Whatever the scale of change, these explanations have given confidence in perusing new ideas that would achieve performance with a limited (or at least considered) chance of failure.

Science provides a framework for exploring the laws and forces of nature. It provides a method of investigation from which explanations of the natural world can be verified. These explanations are useful in explaining the behaviour of construction and the output of research will inform building codes, standards and design guides. The challenge for everyone involved in analysis and choice is being able to stay abreast of the changes to prevent failure and ensure an expected level of performance.

The approach to analysis

Analysis will be based on visualising the dynamic actions and processes involved. This can be greatly assisted by diagrams, mathematical modelling and simulation on the one hand and by experience on the other. Both will require verbal articulation and discussion to ensure understanding. The aim should be to build a picture in the mind's eye of what the solution will physically look like (the suggestion), and the actions required to determine the response that describes the potential failure under working conditions (the analysis). Digital tools will be crucial in ensuring that the (virtual) solution satisfies given performance criteria. Providing a description of both physical construction and then the response to the construction against certain criteria, which if not satisfied would result in the construction being deemed to have failed, is the underlying approach used in this text.

Linking the Design Concept to Construction Choice: Chapter 3 AT A GLANCE

What? The link between design decisions and the choice of construction is fundamental to the realisation of a sustainable built environment. It is also crucial to how the building can be assembled, maintained and subsequently disassembled safely, cost effectively and without wasting materials and energy. This involves an understanding of flows and transfers, general building forms and appearance, and how uncertainty and risk impact decisions.

Why? The link between the design concept and construction choice is arguably one of the most important parts of any building project. It is here that the performance requirements set out in the project brief need to be met to ensure a building that performs as intended and resists failure. General building forms and construction typologies will be fundamental to informing the initial suggestions and shaping choice.

When? Decisions relating to construction technologies are happening throughout the design process. Initial design proposals will be considering the general form of the building and some of the major decisions about the approaches and materials to be used, be it load-bearing construction or framed, assembled using labour on the site or manufactured offsite, etc. As the design concept becomes more refined and defined, so do the initial thoughts about the technologies that will realise the design. This is an iterative process that relies on constant analysis and questioning of choices.

How? The design team will develop the design from the client's project brief. As the design becomes more detailed there will be the need to include specialists to assist with analysis and contribute to the decision making that informs choices. The choices made will need to be implemented by constructors safely and sustainably. This collaborative effort will be informed by digital modelling and simulation tools.

3 Linking the Design Concept to Construction Choice

Appearance is a major aspect of the building design, associated with evoking a response to the building, which is conditioned by social and cultural norms. It concerns the generation of space as well as the creation of physical surfaces and appearance. Appearance can be associated with the creation of style and image as well as aesthetic consideration. This makes demands not only on the choice of materials and details, but also on the space that is available for the construction. There must be a way to link the design concept that develops from the client's brief to the process of technical choice of the elements, assemblies and components. This is necessary to ensure that the design concept and design intention are carried through to the specification and detailing of the building design. The design must determine not only the arrangements and interrelationships of the spaces, but also the flows and transfers required between the spaces. This will allow the function of the parts to be established as a function of the building. Initially the design will most likely be developed with some ideas about the broad options, from which the specification and detail could be developed to ensure the full design intentions are achieved. Good designs require both physical performance and qualities of appearance to be satisfied.

3.1 Flows and transfers

When seeking a link to technological choice, it is necessary to be concerned not only with spatial arrangements but also with flows and transfers in, out and around the spaces as well as through the construction. Building designs must establish an arrangement of spaces, both internal and external, and a way of defining this series of flows and transfers between them. It is by establishing the flows and transfers that the dynamic behaviour of the construction can be visualised. This allows an initial suggestion of the generic forms of construction that may be appropriate for the context. However, perhaps most important, it allows for the identification of the function of the parts of the building and their contribution to the function of the building as one entity.

Flows and transfers take place in one of three ways. The first is in and around the building. Perhaps the most obvious is the flow of people and the transfer of objects. This is of direct concern to aspects of the design that considers the arrangement of usable space. This must be extended to considering emergencies, such as the flow of people when evacuating the building in case of a fire. However, movements in and around the building will also involve air carrying moisture, heat and odour. As a process of natural ventilation this

Barry's Construction Technology: Analysis and Choice, Third Edition. Stephen Emmitt and Tony Bryan.

flow must be planned and may require openings other than those needed for the movement of people and objects. The second way flows and transfers take place is through the construction fabric. This is of direct concern to the choice of structure and enclosure elements. Although the construction is often characterised as passive, the behaviour of the fabric is dynamic in response to these flows and transfers. There are energy flows associated with creating environments, normally heat, light and sound, as well as transfers of forces through structural elements. These flows also include the physical transfer of gases and liquids (moisture) through cracks and permeable materials. The third way is through building services. These are purpose-designed systems that are characterised as active because it is their function to transfer commodities around the building. They may be carrying energy either in a medium such as heat in water or directly with electricity in wires. Water and drainage systems have the function of transferring clean water to points such as sinks and basins and then disposing of the soiled water safely. Signals and data must be transferred in digital communication systems. As active systems they need control. They need to be turned on and off, and up and down, either manually or via the use of smart technologies. This distinguishes them from the passive fabric, where flows and transfers are determined by the properties of the materials and size of the components.

The size and interrelationship of spaces, the fabric of the building and the building services and the interchange and transfers between them must be incorporated into the design concept. The design intention must be made clear so that it can be taken right through the process of choice of the technical solutions. This is important because the people making the conceptual design decisions usually differ from those making the technical decisions, and those constructing the building.

3.2 Concept design

It is widely acknowledged that the most crucial stage in achieving a sustainable, resilient and circular built environment is the design phase. Environmentally sustainable objectives must be incorporated into the very form of the building at the conceptual design stage. All choices about building form, appearance and internal layout that are considered at this early stage have implications for technical solutions; achieving sustainability objectives is no different. It is at the conceptual and subsequently the detailed design stages that decisions are made that impact the choice of building materials and components. The thinking of all members of the design team must be aligned with sustainability and circularity principles so that it underpins analysis and choice. This requires a collaborative approach by individuals and organisations that share the same (ecological) values, principles and ethics. This necessitates an understanding of group decision making and the use of the most appropriate digital tools to facilitate the choice of building materials and components that contribute to circularity.

Architectural, structural and environmental considerations will be embedded in the conceptual design, which can identify general construction forms and systems early in the design process. These can be worked up into specific solutions later. Designs are made reality by technical solutions that can be conceived with the resources to put them into practice. Inevitably most designs are worked out with practical technical solutions in mind. This does not mean that complete detailed solutions are required. It is more likely to be associated with the identification of broad technical options, indicating the materials and sizes, some details and how they would work with the other elements of the building. In some cases the demands of the client may lead to designs that cannot be achieved with immediately identifiable technology, leading to innovative suggestions. This requires a deeper exploration of the general forms and generic details as a reference base for creative solutions.

All buildings pose challenges somewhere between copying the technology and solutions from the last building, to developing new and novel solutions for general forms. Perhaps the key understanding in this process is just how difficult it is to achieve a technical solution to an evolving design. There is no direct relationship between novel designs and technological difficulty. The radical design may, for example, be easily achieved with existing technical knowledge and experience.

Physical performance and appearance

Conceptual designs must allow for the function of each part of the building to be aligned with the level of performance required if the building is to function as an entity. Concepts of appearance involve aesthetic judgements and rely on responses conditioned by cultural and social influences. Appearance should not be interpreted too narrowly for this aspect of design. It involves not only the finish of the surfaces that can be seen and touched, but also the sense of space and the appearance of the space to the beholder. It is as much to do with the perceived feel of the space defined by the building as it is by the look of the building.

The business of providing a functional building is a matter of physical performance. Concepts must be based predominantly on the realities of the physical world and the laws of nature, although the arrangement of physical space influences our psychological wellbeing. This involves physical, environmental and structural judgements, which draw on knowledge from ergonomics, the sciences and engineering. They require an understanding of the physical behaviour of the construction when subjected to the forces of nature. Most definitions of 'good design' would argue that both performance and appearance should be satisfied to attribute any judgement of quality to a building. The response to aspects of the appearance is subjective and will be based on a human, cultural and spiritual responses expressed as emotions. The response to the physical performance will be physiological and psychological, expressed in feelings of comfort and safety. This is also a subjective response, but one that can be measured through surveys of building occupants' perception of, for example, thermal comfort and wellbeing, which can then be compared to building conditions, such as temperature, humidity and airflow.

The design will have to identify not only the function but also the performance level. For example, the environmental design will determine whether a wall will have a function to resist the passage of heat. The design will have to determine the required internal condition and then will have to suggest the relationship of heat inputs to retention rates and the role of services and passive construction in maintaining the design conditions. Given concerns for energy saving, most designs look for retention of heat to balance low energy inputs, which means a high-performance requirement from the wall in resisting the passage of heat. Having established the basis of the function and performance levels of the elements, assemblies and components of the building, it is now possible to undertake the analysis necessary to evaluate proposed construction specifications and details.

Appearance – attributes and response

It is necessary to think about the attributes that will influence the appearance of both the space and actual physical presence of construction. There is a need to judge the response of people to a suggested solution. The response will be conditioned by cultural and socialising influences. The response required may be an aesthetic judgement concerning taste and beauty. It may also be required to convey some message about the purpose of the building or the status or beliefs of the owner. These would have been part of the project brief and should be present in the design concept. It is then important to evaluate the extent to which the suggested solution evokes these appropriate responses.

The requirements of appearance will be primarily judged by our response to the experience of the building and the spaces it creates. Buildings divide space: internal and external, public, and private, illuminated and shaded. These not only inform and enhance the function of the building, but also should respond to the context of site and surroundings. These are deep responses and involve judgements with cultural and social overtones. This generates a paradox in that, while they may be the most difficult to articulate objectively, at the time of choosing the solution it is the attribute that is most easily 'measured' by the users of the building. Since this book is predominantly technical, it is not intended to discuss this analysis in any detail. It may, however, be of value to examine one or two of the issues that contribute to the formulation of this design intent from the brief and the context in which the building is to be placed.

The client's brief must be translated into appropriate spaces arranged to mimic the anticipated activities in both scale and interdependence. In achieving this the design intent must encompass such ideas as style, image and aesthetics as well as context. The need to satisfy the requirements of appearance also influences the space available to the construction since it can only occupy the zones between internal space and external form. The design concept may be confounded not because a construction form cannot be conceived and created to perform, but because there is not sufficient space for it to exist without it adversely influencing the aesthetics.

Buildings are experienced not only from inside, associated with the activity of the building, but also by a wider public from the outside. In addition, the view from inside a building can be an extension of the internal space and change the experience of the internal space. Buildings contribute to landscape and cityscapes, in which they may enhance or detract from the sense and feel of the space they occupy. Some aspects, such as orientation to provide access and movement between inside and outside or to provide the best use of the natural environmental conditions such as sunlight are relatively easy to identify, analyse and justify. Other aspects, such as scale, massing and the sympathetic use of materials are often not so easy to identify, particularly in the context of a statutory town planning framework. What one person perceives to be an aesthetically pleasing building may not be the reaction of all users and the local community. Taste is personal. The aim is to try to achieve a more beautiful and resilient built environment that satisfies the project performance parameters and the wider expectations of society.

Physical space for the construction

The bulk or slenderness of the construction has a profound effect on its appearance. The size and scale of the actual fabric of the building and the detailing will have to be in sympathy with the intended style, image and aesthetics as well as being significant to context.

This is true for both the internal spaces and the external appearance of the building and its contribution to its surroundings.

Much of the expression of appearance is about aspects such as line and surface shape as well as colour, texture and light of the actual surface. These are spatial limits to construction as well as definitions of the surfaces themselves. The construction must be contained within a zone that remains when internal space and the external form have been defined.

This limit to the size and arrangement of the construction must be part of the analysis. It may be that to achieve the required level of physical performance in an economic manner the size and shape of components will exceed these limits and intrude into the visual field of the onlooker, interfering with the required appearance. Moreover, these spaces or zones for the physical construction often need to contain several elements of the fabric and services required for the building. Planning for the space for the elements of construction can become as significant as planning for the space for the occupants.

The approach to analysis

Image, aesthetics and response to context are not subject to the same type of analysis as the physical performance expected of the construction solutions. They do, however, influence the choice and must be part of any evaluation of a suggested solution. This analysis is not a dialogue with the laws of nature but a dialogue with the perceptions of taste and cultural norms. Subjectivity may over-ride objectivity in the analysis and choice process. Normally much of the construction is hidden from view within the building fabric, and therefore appearance is not a consideration. Appearance is important for the elements that are visible once the building is completed. In some designs the natural aesthetics of form of the components is used as part of the design concept. In these cases, shape and size and in particular joints and fixings become part of the desired appearance. We can add to this the texture and scent of finishes.

Some designs employ decoration and appendage to create the desired qualities of appearance. These parts of the construction may play no other part in fulfilling function other than appearance. Their justification lies entirely in the contribution they make to the response they invoke, giving value to a particular form of style and aesthetic expression. No less than any other part of construction, choices must be made about the materials, shape and size, plus some means of fixing. This is still a technical challenge because the parts must have sufficient strength and durability to maintain their effect over time. Historically this may have been more associated with craft, where the emergence of design came from experience with the materials, rather than an analysis of the creative arts. This represents a technological tradition where design, detailing and construction were more closely related in the everyday choice of technical solutions for buildings. This is a consideration for professionals involved in the conservation and adaption of heritage buildings.

3.3 General and constructed forms

Elements and components have a limited range of conceptual generic forms based on their action in achieving performance. Detailed construction solutions must be chosen for a specific design and the resources available and are therefore ultimately specific to an individual building and its location. Within each society a range of common building forms and economic construction solutions arise, and which normally form the basis for choice for subsequent construction solutions. These common forms will be drawn from a range of generic types that are then worked into various specific solutions. The generic type and constructed form are two distinct ways of identifying the construction that is adopted for a building. The conceptual generic type is characterised by its arrangements of components and action in achieving performance. The physical constructed solution has defined materials, size and shape of components, joints and fixings. This construction solution becomes specific to an individual building, although the building solution may be repeated many times, and the details may work in many other types of buildings and be repeated also. The challenge is not to repeat these 'standard' solutions without questioning their appropriateness for a changing climate and changing expectations of society.

For each of these building forms there will be an interrelated set of construction solutions to cope with the size of the building and the resources affordable for each. Construction solutions are usually derived from a limited set of generic types or typologies. Technological choice often starts with the selection of an appropriate general form that is then developed into a detailed construction solution for a specific building and its site. The ability to describe a building in these two forms, the general form and constructed form, at the early concept stage in the design of a building is most important. It should be the big idea from which the individual details can be developed so they work together to ensure the performance of the building as an entity.

3.4 The emergence of general building forms

It appears that the prime determinant of the construction of a building is related to space. The users' demands will translate fundamentally into spaces and their size and interrelationships will suggest the scale and shape of the building. At the domestic scale small cellular spaces lock together, while many commercial or public buildings require more voluminous spaces, often with smaller associated spaces such as toilets, to form the building as a working unit. Design solutions are directly related to this demand for space. Most significantly it influences the structure, but these choices have inevitable consequences on the solutions available for the construction of the enclosure and other elements of the building. Currently there is a wide range of spatial requirements both on plan and in the height required that leads to the use of a range of structural forms. Three general groups of structural form can be identified:

- Loadbearing walls
- Skeletal frames
- Long-span roofs

Each will create a range of building forms, some examples of which are shown in Figure 3.1.

Each building form will be made up of a combination of a range of structural elements. An example of a generic structural element is the beam. With its structural action in bending it appears in cellular loadbearing construction as joists for the floor and a lintel over an

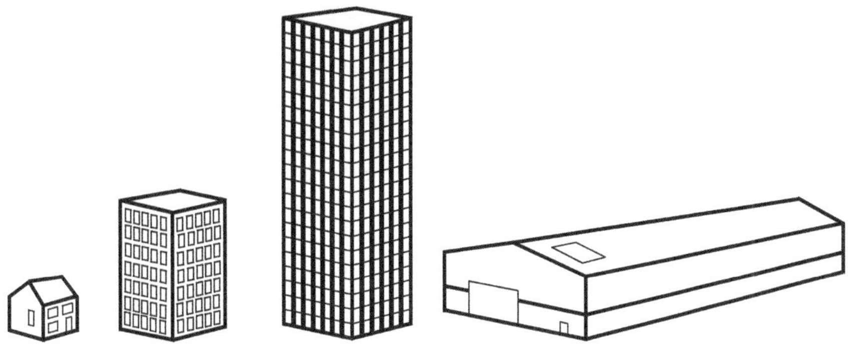

Figure 3.1 Typical form associated with types of structure.

opening. In skeletal frames it appears as a main structural member, and in the long-span roof in a portal frame. This generic type has been adapted for specific solutions using different materials to achieve the required characteristics: great strength, small deflections, small section size, low weight, good looks, low cost, fast erection. Each of these characteristics will have a greater or lesser significance in a specific building. In a house the joist will most likely be timber; in the skeletal frame the beam will generally be steel or reinforced concrete. The portal frame is most likely to be made from steel, but for a building of the right use, scale and size it could be made from reinforced concrete or from timber in a laminated form.

While the ability to enclose the space is the primary determinant of the generic form, both design and resource conditions will lead to a dominant economic detailed solution. This leads to two ways of thinking about a building, each having its own vocabulary to describe the solution. Early in the selection of the construction it is helpful to think about the elements of the building in generic terms. This can then be developed into a full description of the constructed form, including specification of materials, joints and fixings.

Generic elemental – domestic, loadbearing walls

One widespread use for the loadbearing wall form is the house. Described in their elemental form, most houses built in the UK at the beginning of the twenty-first century employ a semi-permeable (cavity) construction for all external walls to provide (with internal partition walls) a stable dry box. This box will be roofed with a tiled covering supported on a trussed rafter structure taking the loads directly to the external walls. This means that any internal loadbearing walls need not extend to roof level. Suspended timber floors will be joisted, often taking advantage of internal walls to limit spans, making some of the ground-floor internal partition walls loadbearing. Ground floors may be ground supported or be suspended floors depending on the site and ground conditions. Foundations are most likely to be strip footings unless the ground conditions are very poor, as shown in Figure 3.2.

Many houses built in the first half of the twentieth century and still in use were not constructed like this. Many have solid permeable masonry walls and have roofs made from individual structural members (rafters, purlins, struts and hangers) which, when acting as one entity, require internal wall support. Later in the twentieth century the width of

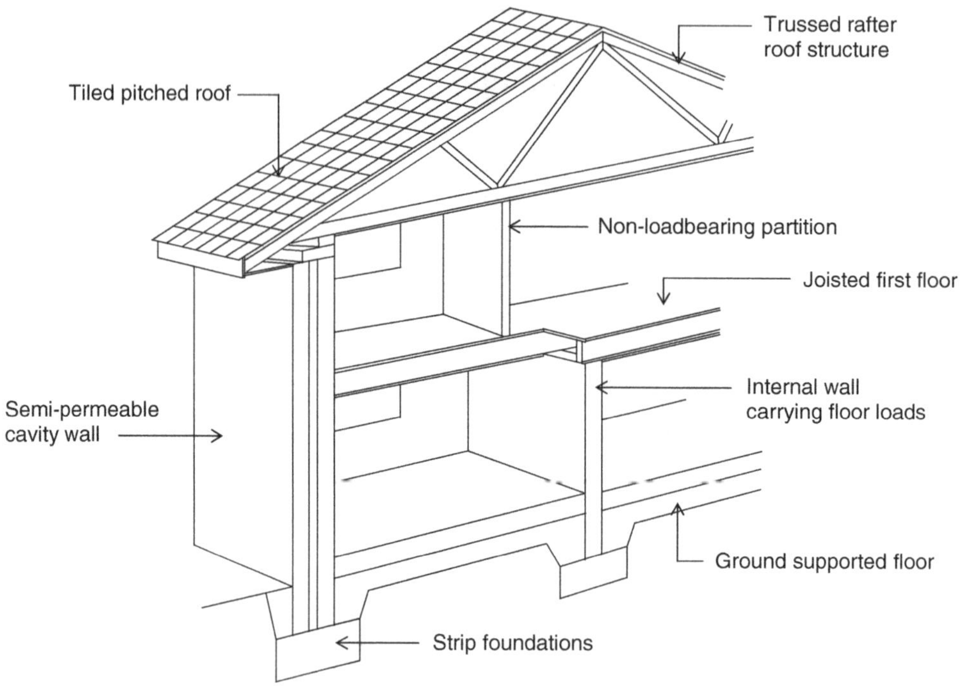

Figure 3.2 Generic description of house construction.

terraced houses reduced and cross wall construction, where only walls between the houses are loadbearing, became an option. This lost the stabilising effect of the box but allowed a range of less expensive external back and front walls to be chosen.

As the loadbearing wall solution is developed into multi-storey flats other elemental forms need to be employed. Floors with greater sound and fire performance requirements become more economical as concrete slabs, even though the spans may not have changed significantly. The loadbearing walls may become prefabricated, taking advantage of the repetitions floor by floor to gain a production advantage. As this arrangement is developed into multi-storey blocks, the alternative structural form of the skeletal frames will become an economic alternative, even though the plan layout still shows walls dividing the building into spaces with limited floor spans. This also allows the possibility of prefabricated pods, such as bathroom units, to be slotted into the frame and connected to services.

Constructed form – domestic, loadbearing walls

None of the generic descriptions talk about the material, the shape, or size of the components, or the joints and fixings employed. They give the elements general names and describe them in terms of their general form or how they perform (Figure 3.2). For the constructed form the details must be worked out to ensure economic sizes and manufacturing techniques. From this a recognisable physical constructed form will appear. The constructed form of the generic elements shown in Figure 3.3. This shows concrete single-lap interlocking roof tiles supported on timber battens over felt fixed to timber-trussed rafters

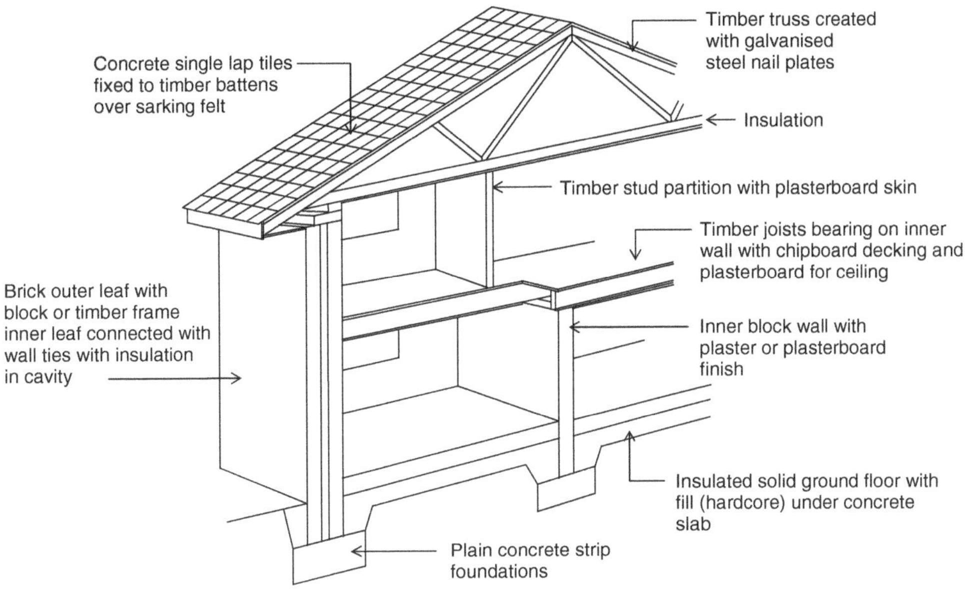

Figure 3.3 Constructed form description of house construction.

jointed with galvanised steel gang-nail plates. The external walls are masonry, although timber frame is a viable alternative. The wall may be constructed of brick outer skin and block or timber framed panels for the inner leaf. This form of construction will not meet the current requirements to limit the passage of heat through the wall. Thermal insulation is required. The most cost-effective way to limit heat loss from inside the building in the winter condition is by incorporation of thermal insulation into the wall assemblage. In summer the design condition is to limit heat gain from outside to inside. As this gain is predominantly from direct radiant heat from the sun, shading or thermal capacity (not shown in Figure 3.3) may be better actions than insulation.

The floors are formed of timber joist supported by resting either directly on the inner leaf or on galvanised steel joist hangers at regular intervals. The floor will be finished with chipboard decking to create a level and stable floor surface. The ground-floor partition walls carry the joists, and they are constructed with loadbearing concrete blocks. However, because the rafters transfer load to the external walls the first-floor partitions only need to be constructed of timber stud with a plasterboard finish, if they can provide appropriate levels of sound insulation. Plasterboard has been used for the ceilings and as a finish for the downstairs blockwork. The foundation would usually be a strip of plain concrete, depending on the ground conditions.

The size of each of these components must be chosen. Common components, such as the brick and block, or the standard range of windows, have been developed in appropriate sizes to work with domestic scale construction and performance levels. The size of some components, such as the trussed rafters and floor joists, must be designed to the size and use of the building. The width of the masonry cavity wall will be determined by the thickness of the thermal insulation required to meet prevailing building regulations. As the requirements become ever more stringent, the depth of the thermal insulation must increase to provide the required thermal resistance. The alternative is to use a different thermal insulation product that can offer the same performance but is less thick. By establishing the thermal performance of the wall it is then possible to establish its thickness depending on the materials and products to be used to satisfy the performance requirements.

In the same way that houses built in the past have been different from the construction shown in Figures 3.2 and 3.3, there is no reason to suspect that either generic forms or the physical solutions will endure into the future. They will evolve and develop to meet new design needs and resource conditions, as explored in Part 2 of this book.

Other general forms and current practice

Buildings using skeletal frames and long-span roofs show a greater variety of both generic elemental types and physical construction solutions. A few examples serve to introduce these forms as both generic and physical entities. The skeletal frame structure has beams, columns, slabs and some means of ensuring overall stability. This basic format can be repeated to generate large buildings both on plan and in height. These structures will, however, need internal structural columns or walls, thus imposing some limits on clear internal space. This structural form has no current rivals in its ability to achieve great heights with relatively open internal spaces. The physical constructed form is generated in either steel, engineered timber or reinforced concrete. The reinforced concrete can be either precast or be cast *in situ*. Most steel solutions have steel beams and columns but concrete floors and either concrete or steel wind-stability provision. This is referred to as hybrid construction as it incorporates the best qualities of two different materials. Reinforced concrete structures normally have reinforced concrete members throughout, although the development of flat slabs as spanning structures has eliminated the need for beams. Engineered timber may also include steel beams depending on the design requirements.

Long-span roof structures are used for industrial, retail, storage, exhibitions, sports venues and other activities that need large floor space under the roof. They are normally only single storey, although they may require considerable internal clear height. Some buildings used for these purposes are not large, and so although termed long span because of their potential the generic forms may be used for buildings of modest scale. Long-span roofs have perhaps the greatest variety of generic elemental forms to be developed into construction forms. Both plane and three-dimensional structures in flat and curved forms have been developed in steel and reinforced concrete and even timber and aluminium.

Both skeletal and long-span structures must be enclosed. The enclosing fabric can normally be supported and restrained by the structure and there is a wide range of options for the external wall fabric. Again, generic forms such as claddings or facings can be identified, with the constructed solution in materials such as metal, concrete, glass, or brickwork. Different structural forms tend to use different enclosure solutions. In long-span structures the building often requires fewer openings, and reducing the weight of the fabric allows economic gains to be made in the structure. This leads to sheet cladding in metals with heavier forms such as brickwork used only around the building at ground level, where it can be supported directly on the ground. In the multi-storey skeletal form the interaction of the enclosure and the structure plus the need for performance often lead to more complex construction.

Building services

To illustrate the idea of the general form and the specific solution, descriptions have been given for the construction of the structure and enclosure elements. Buildings also include a group of technologies collectively known as building services. It is not the intention of this chapter to introduce their general forms and possible solutions. This is not because they are any less significant in supporting the everyday use of the building and the wellbeing of the occupants. Each building will have to be resolved with a mix of structure, enclosure and service systems to meet the performance expectations of the building. Hopefully in this chapter the point about general and specific forms has been made with the examples of structure and enclosure. The general forms and possible solutions for services will be introduced in later chapters in the book.

3.5 Emerging technologies – uncertainty and risk

Initiatives to reduce environmental impact and to make the built environment more resilient will give rise to the need for new technical solutions, both high and low tech. It also prompts us to explore some largely forgotten vernacular building solutions that may offer low tech options to be used alongside modern materials and technologies. Many new solutions will be based on existing technologies and knowledge applied in new ways, but some may give rise to new materials, new technologies and new ways of construction that were previously unknown. These will need to be evaluated not just for their sustainability credentials but also for their performance, cost and production potential over a long period of use and reuse. In this book the issues of sustainability are introduced as part of the evaluation of the design, the physical behaviour and manufacture and assembly of a building. This needs to be seen as part of a circular economy that is dependent on collaboration and the sharing of knowledge to better inform construction choice. As we learn more about the issues and devise solutions that prove themselves in practice, the new constructed forms become the precedent for further evaluation and development. Resilient, net-zero buildings will become the established response in our design and production procedures.

New materials and technologies are exciting and provide opportunities for doing things differently. They do, however, represent a degree of uncertainty and hence risk for the professionals involved in analysis and choice. Particularly in times of change, the process of choice must include analysis that recognises the uncertainties and risks for each building in failing to achieve performance. We must also acknowledge the reality of available resources given the constraints of money and time. When change is slow, there is the opportunity for current detailed solutions to be adopted with little modification from one similar building to another. There is always danger in assuming two buildings to be the same, but if the general form is robust and is not being applied close to a failure cusp the solution can be suggested in detail without much analysis and with little risk. As the pace of change increases, initial choice may deviate from the familiarity of currently detailed and generic forms. The initial suggestion becomes based on an understanding of generic forms that then need to be thought through in detail. Initially this may only need some alterations in materials or detailing, but as demands become more divergent from common practice then alternative elemental generic forms in new combinations may be more appropriate. This is starting to happen as society moves towards a net-zero carbon economy. For the design team the use of AI, machine learning, simulation and generative design software makes the task of producing and analysing potential solutions somewhat easier. It does, however, still require someone to make a choice from the many potential solutions.

In making any technological choice it is important to be aware of the uncertainty and risk in adopting a particular solution. The technical risk is one of failure in performance. As with any risk it must be weighed against the consequences. Consequences that lead to loss of life always require careful attention. Where the consequence is interruptions in the operation of the building, the severity of the loss will have to be evaluated against the risk of failure.

Whatever the pace of change, the choice of construction is based on an understanding of the action of the generic form and the performance of the detailed design and availability of resources to construct the specific solution.

Construction Variables and Choice: Chapter 4 AT A GLANCE

What? In making a choice for the construction it is necessary to specify the materials, shape and size, and the spatial relationship of the parts. Specifying materials starts with identifying appropriate properties for performance. This then must be linked to the availability of the material in an appropriate form with reliable supply within acceptable parameters. Health, safety and environmental considerations are also significant in making the choice. Joints and fixings allow the parts to act as a whole and the assembly process to be achieved. They influence stability, the accommodation of movement and provide continuity of performance in the elements of the building as well as the building as a single entity. Fixings imply a load transfer function, while jointing involves many functions which may, or may not, include a structural role.

Why? The ability to shape materials brings great benefits in using the properties of the materials efficiently in stable configurations, particularly at joints and fixings. Shape is particularly important to appearance and in the fine detail that ensures performance. Shaping, however, requires tools, machinery and skills that must be available at reasonable cost. Size must be specified to define not only overall dimensions but also the required shapes. The gross size is identified by the basic performance requirements, while the coordinating size is determined by the need for the parts to work together. Given the need to join and fit parts together, the coordinating size will have to be modified to a working size that will then need to identify the tolerances to ensure that the detail can be assembled and will perform as intended.

When? Each detail of construction will have to be evaluated for its ability to be assembled within these parameters and still perform. It is not possible to produce perfect construction, so it is necessary to establish what deviation could be allowed or tolerated without loss of performance. In facilitating the process of assembly, the jointing and fixings must be chosen that can accommodate deviations to maintain spatial relationships and the performance of the building. This applies to all construction, whether it is assembled on the site or using offsite production methods.

How? Specifying and controlling spatial relationships can be to a common reference system or by positions relative to other parts of the construction. Given the existence of deviations, if a reference system is used there will be less chance of cumulative deviations, reducing the probability of failure due to dimensional creep. The use of digital design and construction models can help to ensure consistency of choice and avoid clashes between the various parts of the construction.

4 Construction Variables and Choice

This chapter explores the variables that define construction: materials, shape, size and spatial relationships. It introduces the reality of deviations in assessing whether the components will fit together and perform as intended. The chapter concludes with a section on the specific aspects of jointing and fixing as the process by which the parts of the building can contribute to its behaviour as an entity.

4.1 The variables of construction – the outcome of choice

It is important to understand what aspects of the physical construction can be varied to achieve the desired performance of the building. Identifying what can be varied also identifies what must be chosen and specified if an unambiguous solution is to be defined. Many construction materials and fixings are hidden from view when the building is complete. Therefore, appearance may not be the most important criterion for deciding what to use. There are, however, a lot of materials and components that are visible, or partially visible on completion of the building, and these can make a significant impact on how building users perceive and experience the building. Regardless of appearance, choices must be made for the materials to be formed into components that have a defined shape and size. Each component has then to be given a spatial relationship to the other components, together with the joints and fixings necessary for them to act in unity. A simple identification of the variables of choice is represented in Figure 4.1.

Defining the final physical arrangements of the construction involves a specification for each of the component parts of the building, including any joints and fixings, detailed in assemblies of components to create the final building. This specification will have to include:

- The material of which components are to be made, and the quality of that material
- The shape into which the material must be formed
- The size of each component, giving dimensions sufficient to define the shape
- The relationship of each part either to an absolute position defined by a reference framework or relative to another part of the construction or the surrounding environment

This specification determines the space that the components and their joints will occupy. There may be competition for this space with other parts of the construction and/or the usable space of the building. The choice of construction cannot be confirmed until this has been checked, and it has been established that there is a unique space for each part of the building.

Barry's Construction Technology: Analysis and Choice, Third Edition. Stephen Emmitt and Tony Bryan.

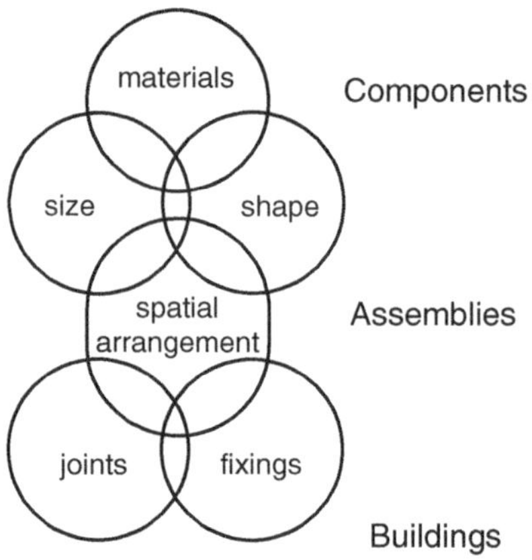

Figure 4.1 Variables or outcome of choice.

Joints and fixings also need to be identified by their component parts, their material, shape and size and spatial relationships, as they themselves are part of the assembly. These assemblies are then connected (where they are often identified as junctions) to form the building. Any components associated with junctions also must be specified. The process of choice is about defining the components, their materials, shape and size and their spatial relationships, including the joints and fixings necessary to complete the building. The focus of technological knowledge is taking decisions about these variables to create a building that is safe and has utility. This is a technical choice that must be made following the creation of a design concept and coming before the management of the production process. Yet it is a choice that must be made in the full knowledge of the design requirements and the available production resources. Understanding how the detailed design will be realised will inform the detailing. This can be explored within digital models to ensure that it can be done safely and cost effectively. Digital models also provide the means to model the disassembly and materials recovery stage, which can also inform the initial choices.

4.2 Choosing materials

Specifying materials is a process of matching the properties of the materials to the performance requirements for the building. We only choose to exploit certain properties when their value is such that, in economic quantities and at sizes and shapes that can be produced, transported and fitted within the space available, they will perform at the required level.

Buildings are perhaps the largest manufactured items widely used throughout all societies in the world. They demand relatively high quantities of materials at reasonable initial cost, and it is therefore not surprising that historically materials used for everyday buildings were those with high bulk, were locally available and with limited process costs. Increasing economic wealth and its attendant higher levels of infrastructure allows the use of materials requiring greater processing off site, and there is less reliance on local materials. It is possible to transport materials around the world to satisfy performance requirements, although this comes with a pollution trail. With economic development and technological innovation there is the possibility for transfer of production activity to factories remote from the site, which in turn gives opportunities to use a wide range of materials. Again, this must be seen in a wider environmental framework with due consideration to carbon footprint and ethical resourcing and manufacture.

The materials that have been dominant in building in the past are timber and stone, with lime, limited amounts of iron for fixings, lead and glass. Fired clay for bricks and tiles has been used for some time in the UK, but up to just before the turn of the nineteenth century these would have been only used locally, particularly in the conurbations where there was a sufficiently large market close to the kiln. In rural areas such as Devon, soil was still being used for agricultural cottages up to the end of the nineteenth century, known as cob construction.

Perhaps the two materials that have made the greatest impact on the size and scale of everyday building have been steel and concrete. The superior strength properties have allowed greater loads to be carried on smaller sections but have introduced the possibility of smaller, efficient spanning members such as beams. This opened the way to the design and construction of large, comparatively voluminous buildings, to a wider group of building clients. In the last half of the twentieth century there was an increasing number of materials introduced into buildings, such as copper and aluminium, particularly for the services. Plastic has made a considerable impact but only where its relatively high cost can be justified by thin-sectioned components. Processing and manufacturing developments have offered some familiar materials new opportunities, with techniques such as laminating timbers and 3D printing of materials.

As well as increasing the range of materials, developments have been made that offer more grades with greater quality control of the more traditional materials, allowing even finer choices to match performance with cost. Evaluating the more recently formulated materials that are new to the construction marketplace is more difficult because many have not yet stood the test of time. While materials are subjected to testing and accelerated weathering in laboratories and via virtual modelling, there is no guarantee that the materials and components will weather as predicted when they are part of a building and interact with other materials. The introduction of additional materials and components with complex assemblies does introduce more possibility for incompatibility, and so even tried and tested materials may fail in new combinations. Adverse reactions between materials and between materials and the environment may threaten durability in ways that have to be imagined before an analysis can be undertaken. This demands scientific research and the development of knowledge from which informed choices can be made.

In imagining failures it is possible to argue that all failures are ultimately materials failures. It is also possible to argue that the failure of the material is always the consequence of a previous sequence of events and therefore never the direct fault of the material. The usefulness of this argument possibly lies in accepting that both are true. In considering the behaviour of the material, it is necessary to identify the significant events in the life of a component when a coincidence of conditions may lead to failure. Understanding the production process as well as the behaviour of the finished building is vital, for the most demanding event in the life of a component may well be in storage, handling, or as part of a partially erected building. An example of this is a precast, reinforced concrete pile where transporting and in particular driving induce entirely different stresses in the materials than when it is under load as part of a foundation.

Materials – health, safety and the environment

While it is the properties of a material that may identify it for use in a building, there are reasons why it may not be preferred or even excluded from the ultimate choice. Initial cost and the availability of local manufacturing techniques have already been introduced, but aspects of health, safety and the environmental impact of the choice must be considered.

There are many reasons why a material may be hazardous to health, particularly during the manufacturing and assembly processes. However, materials have also to be evaluated against their impact on the environment and their contribution to resilient and regenerative development. Society is reassessing its relationship with the natural environment and the demands and stresses we place on it. Much of the concern is associated with our development as a technological society. This new thinking plays a major part in the way we design, build and use buildings. Such concerns as energy use, biodiversity, waste and pollution can be directly affected by our ethical choice of building materials.

In assessing a material for its environmental impact it is necessary to trace its entire history, often known as lifecycle analysis. Any material used in a building will have been part of an ecosystem that will be altered by the removal of the resource we require to make and shape our components. In many cases the effect will be local, and nature will have the capacity to restore the balance in a short amount of time. However, as the volume of development increases and infrastructure becomes available to exploit resources on a large scale, the restorative powers of nature may be severely tested. The impact becomes more than local to the extent that the exploitation of the earth's resources begins to have global implications. Concerns associated with resource depletion, loss of biodiversity and global warming are all affected by decisions to use certain materials in buildings. This is not just in the extraction of materials but also in their local and global transportation. There should be a preference for renewable resources that are ethically acquired and processed locally.

Processing and manufacturing use energy and water and produce waste and pollution. To compare the impact of producing materials we use the idea of embodied energy or embodied carbon. It is the sum of the energy or carbon used to extract and process materials normally calculated up to the delivery to the site. The energy or carbon used in manufacturing may vary even if they are made of the same material. Assessment of embodied energy or carbon allows not only comparisons between materials but also the evaluation of using recycled material where process energy may be less than that of the original conversion. Waste and pollution also pose a major problem in that returning the by-products of processing to nature may place a burden on the environment it is not capable of dealing with. Disposal strategies are a major concern in evaluating the environmental impact of the choice of materials. Disposal is of concern not only at the stage of manufacture and assembly but also at the end of the life of the building. Specifying materials with opportunities for reuse and recycling can help to limit disposal and pollution costs, and this will form another part of the analysis of the environmental impact of the choice of construction.

4.3 Choosing shape

The shaping of components can bring great benefits in terms of performance. It can give strength by corrugating comparatively thin materials and can provide ornament and finish to satisfy appearance requirements. It has a particular role to play at joints and at points of fixing, where shaping such as grooves and rebates can be the major contribution to achieving function and performance. Shaping is, however, dependent on manufacturing processes that require skills, tools and machinery. This adds cost to the component that must be justified by its utility, or a saving in material. It is also dependent on the material chosen, not only for its ability to be worked with current process skills and technology, but also for its ability to be worked to fine enough deviations (precision) to achieve the allowable tolerances required for performance over time.

Early construction methods exhibit the characteristic of minimal working or shaping of the raw materials. Exploiting shape comes with increased resources where skills and manufacturing technique become available and cost effective. Shaping is achieved either by forming the shape directly from a formless material or by working from simple shapes available from the natural form of the raw material. In some cases these are combined where a crude shape is formed and then refined by removing material to work towards the final form. Most forming of materials relies on the removal of material via cutting, which needs to be reused to prevent waste. Techniques such as 3D printing mostly rely on building up the shape via a series of layers of materials.

Shaping must be related to an appropriate manufacturing process. Techniques such as casting, rolling, extruding, or printing, can shape a formless material such as concrete, clay or hot metal. Cutting or machining can then be used to refine or modify the shape. Parts can be added by fixing additional pieces to make complex shapes from simpler forms. Simple forms can also be bent, pressed, beaten or forged into more complex shapes. These processes introduce a range of materials properties associated with workability. They may make a material with all the properties for performance unusable or at least too expensive to use. Many of these processes are essentially factory based; the range of methods that can be undertaken on site is still limited by individual skill and hand tools, leaving pouring, cutting and attaching with some bending as the predominant shaping processes.

Choosing shape and appreciating its performance value and production cost is a major consideration in technological choice. It has particular significance in the issues of deviations and fit, to be discussed later in this chapter. Where parts with complex shapes must fit together, there will be specific demands on accuracy that limit the ability of the assembly to accommodate distortions. Hence, while shaping of individual components leads to many opportunities in jointing and fixing, this must be considered carefully in terms of the components' relationship to the assembly of which they form a part.

4.4 Choosing size

So far it has been established that the level of performance required from the construction leads to the initial choice of material based on a match between the function and a property that can be exploited economically. It has also been established that for each material certain processes, such as cutting or shaping, may be available to make best use of the material and facilitate assemblies that function as whole elements of the building. It is now necessary to establish dimensions sufficient to define the shape, allow manufacture and ensure safe assembly. In the process of choice it may be that there is a limit to the size of the component that will determine the choice of material and/or the shape. This limit may be associated with the need to fit together with other components, or there may be manufacturing/

handling limitations. It may be that visually the size of the component confounds the design vision. It is helpful to see the establishing of dimensions in three phases working towards a manufacturing or working specification:

- Gross size – based on performance
- Coordinated size – based on incorporation with other components
- Working size – based on jointing, fixing, deviations and movement

The gross size will be determined by performance. It represents a minimum size for waterproof materials to keep out the rain or for structural elements to support the loads. It will, therefore, identify the scale and proportion to the solution and may even show at an early stage that a solution is not possible with that material with the chosen shaping in the space available for the construction. If the gross size appears to be sensible to achieve performance economically and within the space available, it is then necessary to look at the dimensions in terms of the detail and how the parts of the construction will go together. This is common, for instance, where the major component is incorporated into brick or blockwork. This is illustrated in Figure 4.2 for the sizing of the depth of a concrete lintel over an opening in brickwork. Structural calculations may show that the lintel must be 125 mm deep. However, the coordinating depth of a brick course is 75 mm (65 mm for the brick and 10 mm for the joint), so the lintel would need a coordinated size of 150 mm.

Establishing coordinating sizes becomes particularly significant in more industrialised and prefabricated buildings that are often designed within a dimensional discipline around

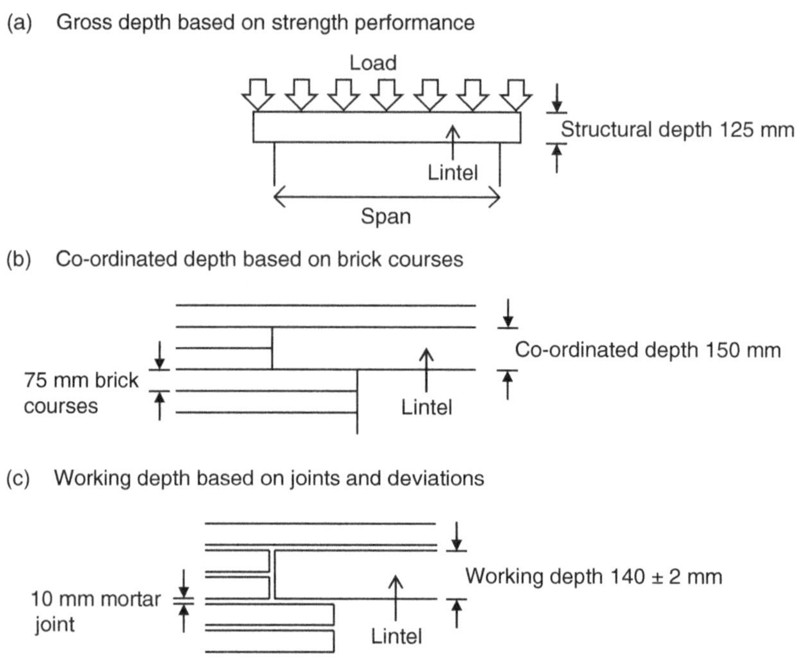

Figure 4.2 Sizing depth of lintel.

a module, leading to a basic or modular size being established. The most common module is likely to be 300 mm. When selecting factory-made components available from manufacturers' catalogues, the economics of production means that they are often only available in a limited number of sizes. Although these will be based on a module to maximise their potential for integration with other components, they do limit the available choice. Bespoke sizes tend to incur additional costs.

The coordinating size does not consider:

- The need for space for joints and fixings
- Allowance for deviations and movement

These have to be considered together, as in many cases the joints will have to allow for the deviations and movement, and the fixings will have to accommodate both deviations and movements if the assembly is to fit together on site and maintain its integrity in operation. Consideration of these aspects will lead to a working or manufacturing size.

Joints are a necessary part of the assembly process to ensure fit in that they can take up deviations in components. Fixings are required to hold assemblies together and these also take up space. Sometimes the joint is also the fixing. The material of the joint and the fixings dictates their own dimensions, with its own deviations. If the fixing is in addition to the joint and this cannot be accommodated completely within the component or in the joint, space must be provided for the fixing. This will have to include any access required to complete the fixing on site. In some cases thought must be given to the space necessary to remove the fixing for repair or routine maintenance.

In the example of the lintel the joints in the brickwork are 10 mm. This will be a mortar joint that will allow for deviations in the brick. It is now possible to establish the working size for the lintel at 140 mm. For a full specification of the depth it is necessary to be clear that in the manufacture of the lintel any deviation from the 140 mm will have to be limited as these also have to be accommodated in the 10 mm mortar joint. Figure 4.2 indicates an allowable deviation or tolerance of ±2 mm. It is now necessary to ensure that the material can be manufactured or worked within these deviations or there is a risk that the lintel will not fit with the brickwork. Working sizes also must consider any allowance for dimensional changes that may take place during the life of the component. These are known as movements or inherent deviations, as they are inherent in the materials chosen.

Dimensional deviation and fit

When establishing a working size and thinking about how assemblies might fit together, it is necessary to think about three things:

- The size of the space into which the component must fit
- The size of the component and its position in the space
- The gap that will be formed between components for the joint/fixings

Generally, a component can be said to fit if it can be positioned within the space so that the gaps allow for effective fixing and jointing. During construction all of these factors (the space, the component, its position and the gap) will vary, and therefore they will all need to have tolerances specified to ensure that it can be safely fitted into the required position.

In traditional construction it is the trade processes that control fit. Procedures such as building in and cut to fit allow for dimensions to be set as the work proceeds. It requires judgement as to how much reduction should be made for fixings, jointing materials, gaps, or movements, but these are part of the craft tradition and are reliant on the knowledge and expertise of the tradespeople. It is not a question of analysis *per* se, but one of common practice and judgement. With the use of manufactured components and the development of prefabricated construction, building in and cutting to fit is not an option. There must be some understanding of these deviations as a basis for analysis to decide on final working sizes before manufacture. With elements of the construction being formed from many components this may have to be based on a statistical analysis of the deviations that can be achieved in manufacture and in assembly. This will be closely associated with the detailing of the joint between the manufactured units.

Consideration of the production process within a digital model will be instrumental in ensuring a good fit. However, it must be recognised that materials shaped with the most skilled hands and/or machines will produce components and assemblies that may deviate from the specified figures. They will be within limits dependent upon the type of material, the level of skill and assistance provided by tools and machines plus the conditions under which the individual must work. It is possible to identify these induced deviations or characteristic accuracies from a particular production process. These will be stated by the manufacturer, which informs choice.

The dimensions that are achieved for the space, the components and the joint will vary within a range of deviations, most of which will be close to the target value with a decreasing probability of larger deviations being produced. This probability follows a pattern known as a normal distribution, as shown in Figure 4.3. This tells us that we cannot predict the size of the space, component or gap in advance. We can only predict the probability that it will be within certain limits. This deviation from the specified value is also true for the specification of the quality of the material.

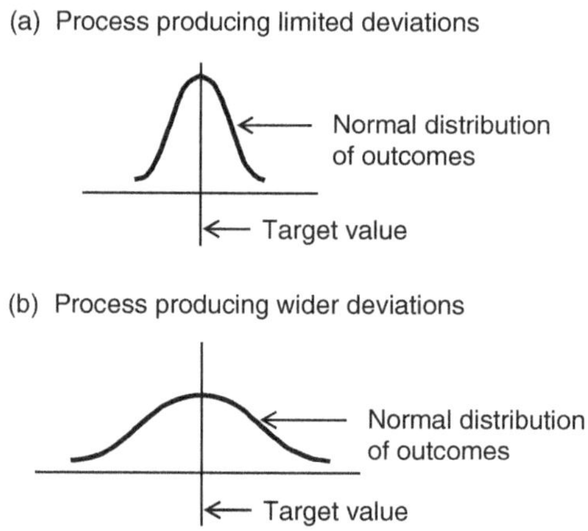

Figure 4.3 Identification of achievable deviations.

Because these production deviations are a reality, it is necessary when evaluating the dimensions of components and assemblies to consider how such deviations will affect the performance. Some limits will have to be placed on these induced deviations, for, if too great, the performance will be impaired. These limits, or allowable deviations, are known as tolerances, or what that assembly can tolerate before performance may be affected.

Tolerances are quoted as a range within which an assembly will successfully perform. They are limits where it is deemed that components installed outside these limits will fail, or at least risk failure in the assembly. They are represented graphically in Figure 4.4. They take the form of a ± value from the specified size. While most specified sizes are set as an average where equal deviations on either side can be tolerated, some specifications are maximums (or minimums) allowing deviations in only one direction.

Knowing the probability distribution for the chosen production process and having an allowable tolerance set by the detailing allows an estimation of the probability of fit. If the allowable tolerance is greater than the deviation that will be achieved for the great majority of the components from the chosen production method, the probability of assemblies failing will be small. It would be usual to ensure that the detailing chosen can tolerate the deviations inherent in current production processes. However, in some designs it may be necessary to consider revised or even new production processes that can achieve the required tolerances, but the cost of this development and the risks involved must be understood before the proposal is adopted.

It is true to say that before a solution can be specified with complete confidence an analysis of both allowable tolerances and inherent deviations (characteristic accuracies) must be made. This will involve an understanding of both the behaviour of the construction under working conditions and the outcomes of the production process before a solution can be said to be a viable proposition against performance criteria. This can be achieved with virtual modelling.

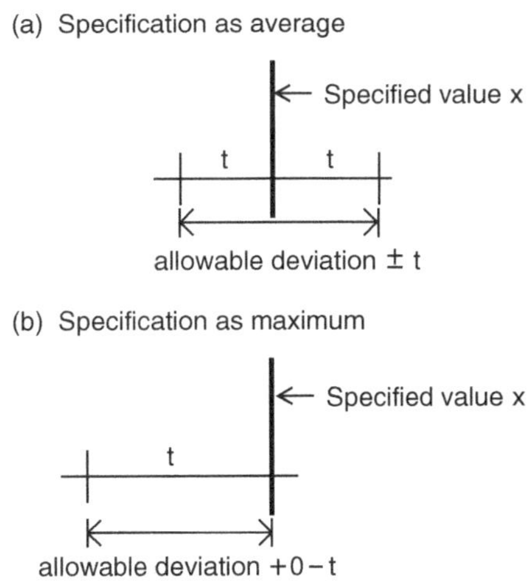

Figure 4.4 Specification of allowable deviations.

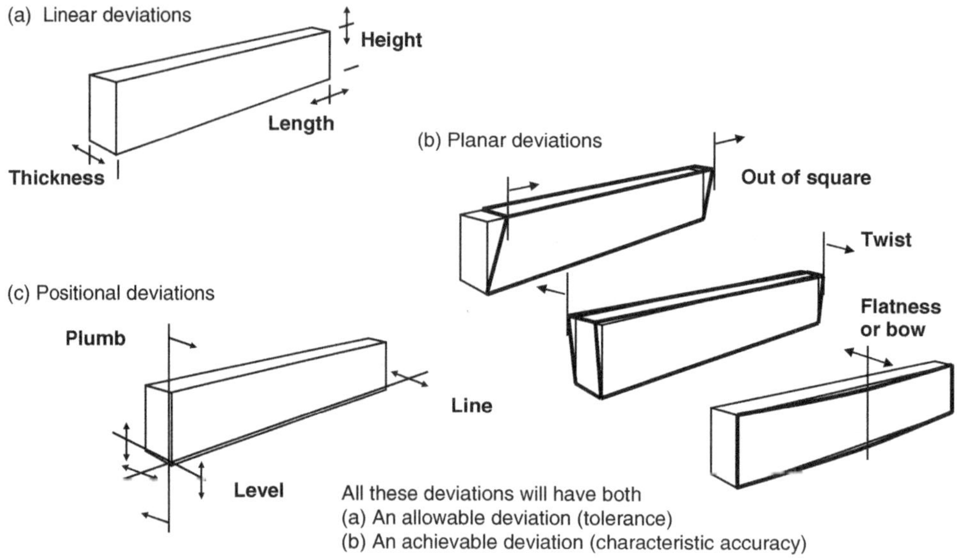

Figure 4.5 Possible dimensional deviations.

Types of dimensional deviations

The previous section looked at the relationship between inherent deviations and the tolerances that can be allowed to ensure fit and operational performance. In defining the component, its space and the gaps that will arise for the joints and fixings, it is important to realise the ways in which deviations can distort as well as just affect size and position. Figure 4.5 illustrates a simple rectangular solid component, but this could also be the space available for the component or the space representing the gap between components for the joints and fixings. Variations in size can be seen as linear deviations affecting the three principal dimensions. However, these may not be the same at all points across the component and this will cause distortions or planar deviations of square, twist or bow. The third set of deviations is positional associated with the line, level and plumb aspects of site control that ensure the spatial relationships are maintained.

4.5 Spatial relationships

There are two ways of defining relationships that can be employed to specify position:

- Referenced – specifying a position in space for each individual part within a common reference framework
- Relative – maintaining a positional relationship with several other parts of the construction

The use of each will depend on the scale of the building and the stage of construction. Since the process of construction is predominantly building one part after another, dimensional deviations can become cumulative if relative positions alone are used. It may be that the fit in parts of the building may become difficult to achieve within their own allowable deviations because specifications did not consider the possibility of knock-on effects within the entire construction. Components can move further and further away from the intended spatial arrangement that the relative dimensions were intended to maintain. This is despite the fact that all previous dimensions have remained within specified deviations.

For construction where cut to fit and building in can be employed, spatial relationships can be controlled to maintain the detailing of the parts and the joints between them. By working within tolerances, parts can be adjusted for the specific circumstances. In this way the joints and fixings can be produced in the required position and within the acceptable limits of their allowable deviations. In modern construction the process is more concerned with the assembly of manufactured components of predetermined size. The preformed parts tend to be built one on top of the other using volumetric construction or positioned within a structural framework. The possibility for dimensional creep or moving outside spatial constraints may be high if too much of the specification is relative. It is, therefore, common to specify most of the early stages of the construction to a reference framework.

The frameworks developed to control position in this way are usually grid lines on plan and finished floor levels on elevation. These are illustrated in Figure 4.6. The 'lines' as they appear on plans and sections are planes that 'box up' space and from which positions can be specified. The control 'box' should not be too small, because the positioning of finishing,

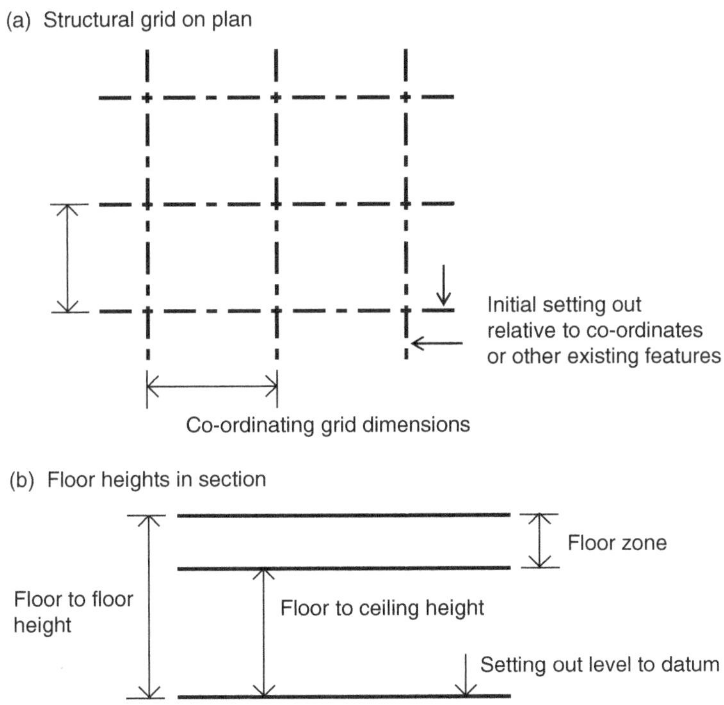

Figure 4.6 Dimensional control framework.

fixtures and fittings can be relative to the major construction so long as they have been well controlled by referenced dimensions. For this reason the grid system chosen is usually associated with the structural layout for the building. Every effort would be made to keep it simple, with repeating dimensions to avoid mistakes in both design and production phases. It is these grids and floor levels that would be the basis of on-site dimensional control. For all but simple building this would usually involve the use of surveying equipment, tapes, levels and theodolites. Advances in this type of equipment including electronic distance measurements particularly associated with theodolites in total station equipment have led to the use of coordinates for site control.

4.6 Joints and fixings

Where two or more components or elements of a building meet, there is a need to consider how they should be joined. Where one or more of these parts requires support or restraint from the others, there is a need to consider how they should be fixed. Joints and fixings are the all-important details that:

- Hold the construction together
- Allow it to be assembled, and disassembled at a future date
- Ensure the performance of the building

Like components, they are made of materials, and have shape and size. They have to work within gaps or their own space between components. They only work within certain limits of that size (allowable tolerances) and can only be produced within limited accuracies (induced deviations). They then have to be positioned within specific limits. They are subject to the same considerations of performance and production criteria as the components they join and fix together.

Joints and fixings facilitate the creation of the whole from a series of parts. They are crucial in the overall behaviour of the building. They influence stability, the ability to accommodate movement and they provide the continuity of performance of the elements of the building of which they form a part. They also play a major part in facilitating the assembly process, particularly where many of the components are manufactured offsite, when the joints and fixings become the focus of the site activity. Buildability, safe working and skill requirements, become a matter of handling the components and the completion of the joints and fixings. There will be the need for maintenance and replacement of elements of the building. The detailing and the fixing should allow for ease of access and safe disassembly of the redundant element and replacement with new. Similarly, at the end of the building's serviceable life the fixings should allow for the safe disassembly of the building without causing unnecessary damage to elements that may be reusable elsewhere.

It is often difficult to separate out what is a joint and what is a fixing. Fixing implies a support or restraint function with load transmission capabilities. Joints tend to describe the whole detail where materials, components or elements come together. The joint may have many functions such as weathering, fire, sound, or movement and may not include a structural requirement that implies the action of a fixing. In some cases components may be fixed independently and are stable but need to be jointed to achieve continuity of function across similar components. The requirement of a joint to isolate movement in a building needs special attention. Fixings may also have to be chosen to allow movement in at least one direction or the joints may have to be created totally independently of the fixings.

Several joints and fixings are shown in Figure 4.7. In traditional timber jointing, such as the mortise and tenon, the detail is entirely based on shaping the components to be joined, which then becomes the fixing of one part to the next. Although glue might have been used, often no other components were necessary. The joining of plaster to a wall also needs to provide fixity. With old plaster systems, laths allowed the wet plaster to form a shape, giving a mechanical key, while modern plasters can achieve fixity with adhesion alone. The jointing of plasterboard to studs is achieved with a mechanical fixing, a nail or a screw.

Joints are often subject to performance requirements other than fixing, such as appearance, weatherproofing or fire resistance. The joint to be made between the sheets of plasterboard in the stud system shown in Figure 4.7 needs to consider movement that may cause cracking of the jointing material, spoiling the appearance. Joints must be reinforced with tape to resist these forces. Cladding panels on the external face of a building are normally fixed to the structural frame independently, being supported and restrained with a range of fixings such as bolts and angles. The joints between panels then function to allow for deviation, both inherent and induced, and provide continuity of appearance, weathering and other functions of the cladding system.

Joints between structural members have a particular significance since the behaviour of the joint determines the distribution of stress in the structural system. The detailing determines whether the joint can rotate or is stiff (fixed). This determines the way the forces are transmitted from one element to another. The function of joints and fixings to facilitate

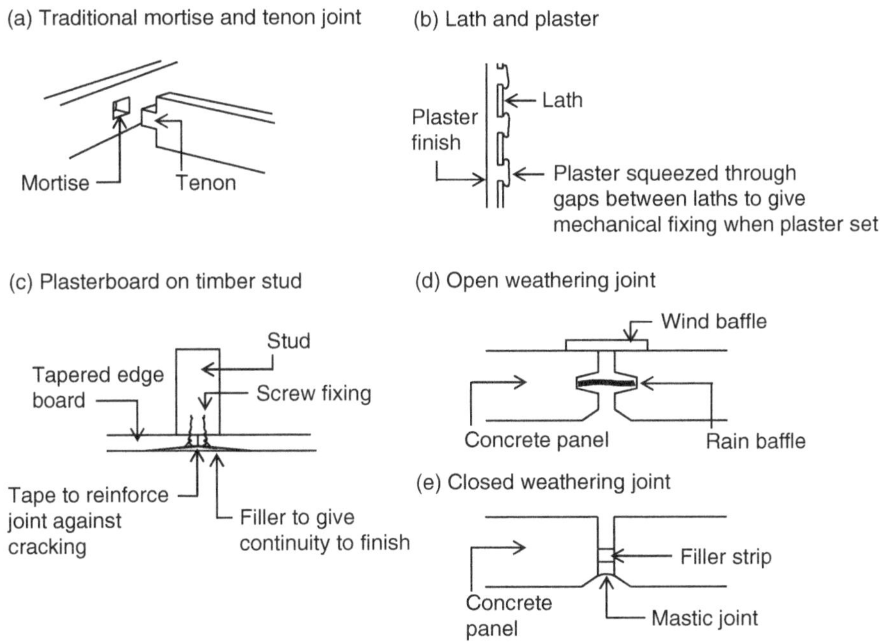

Figure 4.7 Joints and fixings.

assembly needs to consider the issue of deviations, such as deviation caused by changes in temperature or moisture. They must be detailed to work within acceptable deviations, but they can also be used to absorb or even out deviations to maintain the position of components. Joints and fixings that can maintain local allowable deviations ensure the achievement of fit throughout the building.

There is now a wide range of fixings, the exact form being determined by the nature of the forces being transmitted, the part to be fixed and the structure to which it is being attached. Its form will also be determined by the need for accuracy and adjustment in the positioning of the part to be fixed. Options include welding, gluing, nailing, screwing and bolting and may involve brackets or plates. Specifications must include shape, size and materials as well as the pattern or position of fixings. The choice is also influenced by economics, which can be significantly affected by the tools required and the available skills. The choice is also influenced by sustainability, which should push us to the use of non-destructive fixings and fitting to facilitate disassembly and recovery of valuable resources.

Performance and fit depend on the careful consideration of the joints and fixings. Problems are created and resolved around such details. Careful choice and specification of the materials and components must be matched by a similar attention to detail on joints and fixings if the building is to work as a whole during its entire service life.

Defining Conditions: Chapter 5
AT A GLANCE

What? Building is concerned with providing physical construction to create environmental conditions within the building that are modifications of those prevailing externally. These modifications are required to fulfil both physical and social functions. In providing physical construction, the building will have an impact on both the physical and social environments, and these must be understood, be defined and be part of the analysis and choice. Both a target value and an acceptable range must be identified for several parameters to define the internal conditions to ensure a comfortable, healthy and safe environment. These will depend on the activities being carried out in the building.

Why? The activities within the building generate inputs that will modify the internal conditions. If these inputs take these conditions outside the desired range, the building will have to provide the means to maintain the required internal environment. Under operational conditions the interplay between changing external conditions and inputs from the activities and the construction must be seen as a dynamic system: a series of flows seeking equilibrium. In addition to understanding how internal conditions are to be maintained it is important to understand how the choice of solutions will impact the environment in general.

When? The required conditions, expressed as a set of performance requirements, will be established early in the briefing process. They will subsequently be tested during the design process as options are explored and analysed. Designing and detailing the building may reveal the need to reconsider some of the performance parameters and required internal conditions as new knowledge is developed.

How? Understanding the building as a physical and social environment will be instrumental to defining the required conditions and the specified performance levels. Because external conditions influence so many aspects of the analysis of a proposed solution, the same aspect of climate, say rainfall, may have to be defined in several ways depending on which part of the analysis of a solution is being carried out. Aspects of climate are also important when deciding on appropriate methods of manufacture and assembly. Internal elements of construction often separate areas of different activity requiring different internal conditions. In this case the definition of the external environment is another but different set of internal conditions. When modifying environments, the construction has different conditions on either side, which means that the conditions within the construction are neither internal nor external. Identifying what the micro-environment is within the construction and its rate of response to change will have to be analysed.

5 Defining Conditions

This chapter identifies the fundamental importance of the creation of both physical and social conditions to the choice of construction. It establishes the need to specify internal requirements, define external conditions, and then to identify the contribution of the activities of the occupants and the building to the internal conditions. These conditions will be ever changing and therefore the whole system must be seen as being dynamic, essentially a series of flows seeking equilibrium. The chapter concludes by identifying how this process can be applied to the choice of internal elements, such as floors and partitions, and introduces the idea of microclimates within each element of the construction.

5.1 Physical and social conditions

It could be argued that buildings are fundamentally concerned with providing physical artefacts to create the conditions for the wellbeing of those who use them. This statement would raise in most people's minds the modification of the natural environment: from the weather conditions outside the building to a less variable, more comfortable and safer environment inside. This is the primary technical role for the construction, and therefore a clear understanding of the external prevailing conditions and a clear specification of the required internal conditions are of primary importance. These need to be defined before any process of choice can begin.

Increasingly the external environment is being modified by human activity and affected by noise and air pollution. Any understanding of the external environment must be more than just the weather conditions that prevail and how they may be expected to vary with climate change. It becomes necessary to understand the impact that producing and operating a building will have on the local, regional and global environment. The case for the need to understand the interaction of physical conditions is easily understood, but solutions chosen only against criteria of modification of the physical environment are not sufficient. Buildings are created in and for a social context, in which they serve social functions that inform performance requirements. This simple idea is shown in Figure 5.1, which illustrates that the user's interests in modification concern both the physical and social worlds. If an activity can be carried out effectively within the limits of prevailing external conditions, and it is socially acceptable to do so, there will be no requirement for a building in any form.

It is important to remember that buildings have an impact on social conditions with both the internal and external spaces they create. While an understanding of social behaviour is

Barry's Construction Technology: Analysis and Choice, Third Edition. Stephen Emmitt and Tony Bryan.

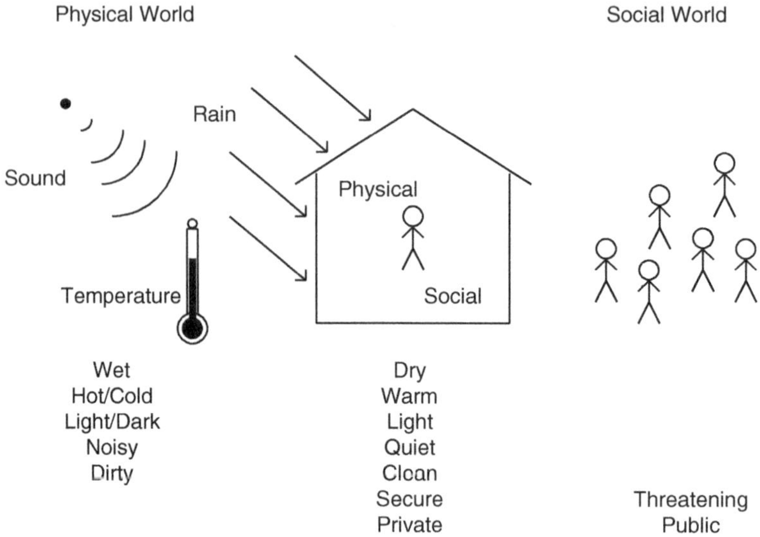

Figure 5.1 Physical and social environments.

beyond the scope of this text, it is well established that building may be part of social problems, but they can also be part of solutions. This is recognised in many aspects of urban design in areas such as crime prevention, community safety and inclusive design. Legislation, building codes and standards and design guides all assist the design team in realising a building within current legislative frameworks of what is deemed to be acceptable and safe. The social context also determines the resources that become available for building. There is a need to understand the influence of the economic, political and cultural environments on the choice of construction set against its cost and its value in its wider social interactions. This has an influence in the specification of the required internal environments. The expectations of the user and the variations in internal environment deemed acceptable are set against these wider social standards and the economic possibility determined by the amount of income clients are prepared to spend on their buildings.

5.2 Activity and construction-modifying environments

Construction modifies the external environment to create the desired internal conditions. However, the activities within the building will themselves contribute to the internal environmental conditions. Cooking generates moisture, heat and smells; machinery generates heat and noise; and cleaning activities generate waste. These activities will modify the internal environment and must be controlled, which may increase the technical demands of the construction.

It also must be appreciated that the construction chosen may modify the environment. This is particularly true of the services in building that often create heat, fumes and noise, but it can also be true of the passive construction. Poorly detailed insulation to resist the passage of heat to provide a thermal environment with lower energy inputs can induce condensation and create damp conditions within the building interior. Finishing materials can give off toxic fumes in the event of a fire. Both are examples of the choice of construction affecting the achievement of the desired internal conditions.

5.3 The dynamics of the system

Although a stable internal environment may be the objective, it needs to be appreciated that the external conditions and the levels of activity within the building will change in a variety of ways. This will produce constantly changing internal conditions. The whole system can be seen as a series of flows and transfers seeking equilibrium, as shown in Figure 5.2. This dynamic makes the process of control of the internal conditions a significant aspect in the process of choice.

Predicting the possibility of exceeding design limits in dynamic systems needs an understanding of both the variability of the conditions and the response of the construction providing the modifying effect. A major consideration in choosing a solution may be its ability to respond in sympathy with the expected magnitude and speed of changes in the environment. This has been recognised, for example, in the advantages of high thermal mass in limiting variations in the internal temperature between day and night, hence reducing the need for energy to heat or cool a building.

To analyse the behaviour of the construction it is necessary to define the following:

- Target internal conditions and acceptable limits within which the activity can operate
- Design limits in the external environment and their pattern and variations
- Inputs to the internal environment from the activity and construction including the response to changes in conditions

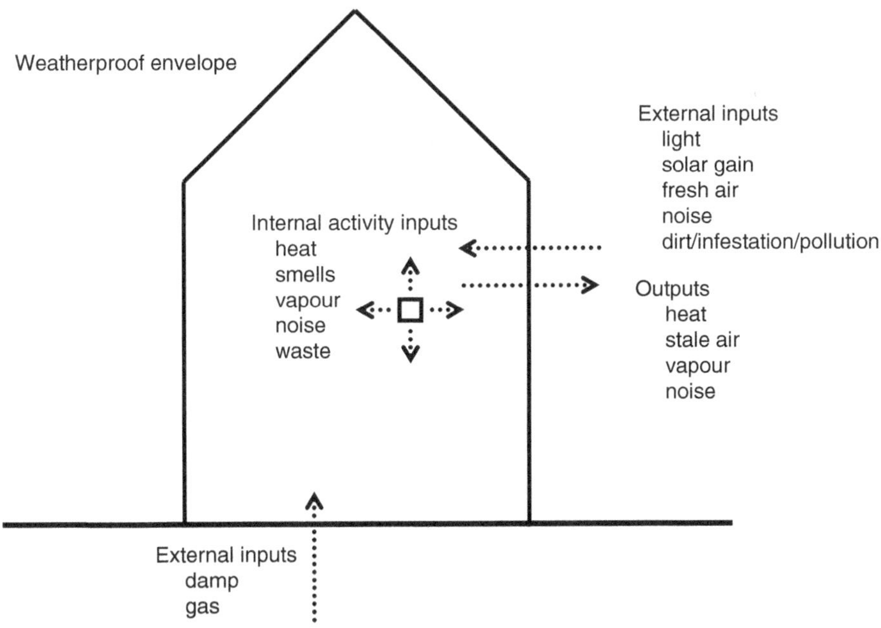

Figure 5.2 Environmental flows – a dynamic system.

Target internal conditions

Defining the target or ideal internal conditions is a question of physiology and psychology in the case of people; engineering in the case of machines; and spoilage processes in the case of stored materials. Aspects of the environment important to the wellbeing of life, machines and materials have much in common. Temperature and dryness are often the most significant to the wellbeing of all three, although ideals and limits may be different. In cold stores the temperature is dictated by the type of stored food. Any people or machines that also must operate in the store have to be protected by other technologies (e.g. thermal clothing in the case of people). In addition to temperature and dryness, consideration must be given to aspects such as humidity, light, noise and air quality. Concerns over air quality, associated with hitherto unrecognised contaminants that occur in very low concentration, such as radon, illustrate how new aspects of environments can come into play as we strive for more demanding and less variable internal conditions.

People, machines and materials do not, however, need precise environmental conditions and can operate within a set of limits. Indeed, in the case of people, constant, unvarying conditions can cause fatigue and some variations are desirable. Specification of internal environments is greatly influenced by social expectations and the resources clients are prepared to devote to the building. With low levels of resources available for the construction, wider limits of environmental conditions may need to be accepted. For example, in the early factories of the industrial revolution, concentration on machine efficiency often made noise and dust levels far from ideal for the workforce. Social expectation is often now embodied in law to ensure clients make resources available to maintain healthy and safe environments within their buildings. Many regulations and advisory standards now set design limits for comfortable, safe and healthy internal environments.

Establishing external design conditions

Establishing external climatic and social conditions is clearly a different task from setting internal conditions. It is necessary to understand the broad patterns of events, with the probability of extreme conditions being identified. The broadest climatic patterns can be identified by global classifications. The fact that the UK lies in a temperate zone gives an overall picture of weather conditions. However, regional, and even local conditions may well determine the pattern of variations and extremes. The south-west of England has the mildest climate and fewest frosts. The highest winds occur along the west coast of Scotland. Variations also exist between urban and rural areas where the pollution and the modifying influence of existing developments can have a significant effect on the external temperatures in cities (known as the urban heat island effect), air quality and noise pollution. All these may influence the orientation of the building on the site and the choice of materials and detailing to ensure the required performance can be achieved. This would also be true of the social conditions, indicated by broad patterns such as inner city, suburban or rural classifications. These may give an indication of the social conditions, but often much more local knowledge will be required if the performance of the building is to be achieved.

The external conditions need to be understood to evaluate most aspects of performance. The need to define climate for environmental control is straightforward: temperature variations and sun paths for heating and cooling requirements, rain and wind conditions for weathering to ensure the building remains dry. However, external conditions also need to be known for structural performance and for durability. Both snow and wind place loads on the building structure but in different ways: sunlight fades colours and embrittles plastics; saturated bricks subjected to frost may disrupt under the internal pressures created during the growth of the ice crystals, leading to damage. It will be noticed from the examples that combinations of climatic factors are often the worst design conditions. Wind and rain can be the most testing conditions for water penetration; snow and wind may give the worst loading conditions; wet weather followed by frost causes damage to brickwork. Defining single factors may not reveal the most testing conditions. The pattern and variations in the climate must be understood well enough to visualise the significant combinations as well as individual extremes.

Establishing variations and extremes requires knowledge of conditions beyond that of the daily cycle or the annual seasons. Extremes do not occur every year. Some of these seem to be identified with long-term cycles or trends and some seem more random and established patterns of climate are changing as the planet warms. Since most buildings have design lives measured in decades, cycles, trends and random extremes that happen over these timescales need to be investigated and accommodated to ensure a durable and resilient building.

For each stage of the analysis (environmental, structural and durability) it is necessary to visualise the most testing set of conditions to maintain performance. For durability this will be dependent on the materials chosen, for the agents of decay are specific to each material. Each set of conditions will then need to be defined in the most appropriate way. Knowledge of annual rainfall gives a broad indication of the magnitude of the problem faced in choosing the construction for a building in that area, but it will not give the whole picture. Occurrence during the year and some measures of intensity are required. Rain that occurs in a single season of the year at high intensity poses different problems from rain that occurs in light showers throughout the year. Intensity over a period measured in minutes will determine the size of the drainage system, while lower intensity over some hours, particularly if associated with high winds, may determine the detailing of weathering to the walls and roof. In another set of circumstances, prolonged light rain may penetrate porous materials sufficiently to cause dampness on the inside of construction well able to resist a sudden downpour.

While this rainfall example is directly related to the physical climate, the principle of visualising the most testing conditions is true for other performance requirements. Changes in social environments, say with crime and terrorism, also need to be established in some way that allows a design solution to maintain performance of the internal environment even with the perceived levels of threat. Crime figures may allow some quantitative analysis of probability, timing (day or night), and mode of entry, which allows choices to be made as to how best to maintain performance.

Because the external conditions are variable, it becomes a design decision as to which limits are to be considered (or given most priority). Generally, design involves establishing the probability of the occurrence and a level of 'value' on the disruption, danger, or cost of the consequence of the building failing to perform. Rare but devastating or life-threatening events may be given similar importance to common occurrences that cause inconvenience or a lowering of operational efficiency. It requires the client to assign value to any loss of performance associated with prevailing social and economic conditions. It may also be subject to the law, for in many cases loss of performance poses a danger that the client may not value as highly as society in general. Over the years this has been deemed to be the case with loss of life in a fire and this has led to extensive fire safety regulations.

Yet another aspect of the analysis of proposed construction that will require knowledge of the external environment is its manufacture and assembly. Conditions influence the accuracy that can be achieved, leading to variations in the achievable deviations. Protection for operatives and the partly completed building work will be influenced by the weather conditions. Efficiency and hence cost will change with conditions through the year. The way the conditions are defined will again vary, depending on the purpose of the analysis and the planned timing of the work. For example, the number of hours of daylight or frost days per month may be the most appropriate way to seek information about the external environment to identify the effects on the decision to specify works to be completed on-site in the winter months. Social conditions also influence decisions on the production process. The labour market is affected by local, regional, national and international factors, which influences the availability of skills and trades, and its cost. Site security will be influenced not only by local crime rates but also by legal and social obligations to maintain a safe site to protect site operatives and the public.

Inputs from the activity

For each aspect of the activity, the types of outputs and the quantity of each must be established. People will produce heat, moisture vapour, other products of exhalation, odour, etc., depending on the activity being carried out. Some of these may be helpful, such as heat gains in winter to maintain temperature, and some will militate against air quality. The contribution of moisture vapour can have a significant influence on the formation of condensation if ventilation rates are not sufficient. Machines can produce noise and heat, and there are concerns over the electromagnetic fields emitted by electronic equipment. While most stored material is inert and contributes nothing to the internal environment, this may not be true if the material stored is organic. For example, the bulk storage of straw can, under some conditions, generate sufficient internal heat to ignite and start a fire.

As with the external environment, broad patterns of these internal contributions may be established, but questions of variations in duration and rates of activity will have to be determined. Designers often look for the single figure to represent inputs from the activity, but if variations are great or change rapidly it may be necessary to analyse the dynamics of the change to match the response of the building more carefully. If data are not available for activities, studies will have to be carried out to determine the rate of net inputs into the environment for appropriate construction solutions to be developed. This data may be available from post-occupancy evaluations of similar buildings with similar types of occupant behaviour. Virtual modelling can also assist with the prediction of internal environmental conditions.

Inputs from the technical solution

Clearly the effects of the construction and the services cannot be defined before the construction and services are chosen and must, therefore, become part of the process of choice. Challenges associated with the building must be foreseen and minimised at the beginning of the process by the broad strategies developed, rather than by resolving them later in the specification and detailing phases, which may result in inefficiencies, poor decision-making, and hence buildings that do not perform as anticipated.

Deciding on broad strategies involves decisions on all the performance requirements and this is therefore a careful balancing act. This includes the balance between passive construction and the active services. With increased demands for quality in the internal environment with smaller variations, active services are difficult to avoid. It is the services that are most likely to generate outputs that affect the internal conditions and use energy. Some input can be used to advantage. Heat recovery from exhaust air can reduce the size of the heating system and reduce overall energy use. However, the noise from the fans necessary to circulate that air may be transmitted through the ducts. This in turn will require decisions on the use of silencers or sound absorbers in the ducts. The materials for these now must be chosen with care as fibres from the sound absorption material may be picked up in the air stream and carried to the rooms the ducting system serves. Designs of building that can take advantage of natural ventilation generated by wind and other pressure differences have considerable advantages, but this must be part of the overall design concept and not only a question of detailing.

Most construction materials are inert and therefore add nothing to environmental conditions. Moisture from wet construction at the early stages of occupation of the building has been largely overcome with the use of dry forms of construction.

5.4 Choice of interior elements of the building

So far, the analysis of internal and external conditions has focused on the inside and outside of the building. However, buildings are rarely used for a single activity. Although there will be an overall purpose for the building, fulfilling that purpose for the user will almost certainly involve a series of interrelated but separate activities. In specifying internal elements, floors and partitions, in the building the notion of an external environment may refer to another internal space used for a separate activity. In this case the internal/external labels are interchangeable.

Problems posed for the construction to solve will depend upon which way the undesirable elements of one environment will affect the second. The task of the construction depends upon which way the flow disturbing the desired conditions will take place. It may become the task of an internal partition to modify sound transmission in one direction and heat in the other if it is between an unheated factory area and an office.

5.5 Understanding what goes on within the building fabric

Although apparently not of immediate interest to the building user, the process of modifications of environments creates intermediate or microclimates within the building and within the building fabric. They can become highly significant to the running cost or life expectancy of the construction, and this is a concern for the building user, operator and/or owner. Perhaps the best example of this is the formation of interstitial condensation within a wall or roof assembly causing damp, when it appears to be dry on both sides of the wall or roof. The wetting of thermal insulation will reduce its effectiveness and, if persistent, may create the condition for rot or corrosion, leading to premature failure of the structure. This can come about through simple day-to-day tasks, for example not opening a window or not using an extract fan when cooking or bathing. While building designers and constructors cannot control the behaviour of the building users, it is possible to install extract fans that operate when humidity levels reach a certain threshold and smart technologies that control humidity levels. The important point to make is that in defining conditions the design team will be influencing how the building performs in use. Thus it is necessary to understand how users may use and misuse their immediate environment. The analysis of these intermediate conditions that rarely reach steady state is another task for the technologist seeking to provide satisfactory performance throughout the range of environmental conditions that may prevail.

Building pathology

The field of building pathology provides a rich source of knowledge for the design team. This body of knowledge has developed through the investigation of building failure, and over the years building designers and engineers have come to better understand the implications of building failure on building performance. By investigating the root cause of a problem it is possible to design buildings to prevent the occurrence of a similar problem in new building projects. For example, our understanding of moisture migration in building structures has become highly sophisticated through testing and research. This knowledge can be used as guiding principles and to inform computational software that can model and simulate physical behaviour of the fabric under a wide range of conditions.

Knowledge gleaned from building pathology can also inform the design team about the performance of relatively new materials and components in addition to reconsidering traditional and vernacular forms of construction. For example, there has been renewed interest in bio-based materials and use of traditional materials such as lime renders instead of cement plasters to better temper the internal environment.

The Resource Base: Chapter 6 AT A GLANCE

What? There are key resources that must be available for the successful creation of a building. These are: materials, knowledge and skill, production equipment, money and time. Materials are the primary resource, being directly related to the quality and performance of the building. Chosen for their properties to fulfil performance, they also must be workable to make components suitable for incorporation into the building. This requires knowledge, skills, production equipment and finance. The decisions are taken in a timeframe and with a background of making a positive impact on the environment.

Why? The level of development in a society determines the ability to process and shape materials as it determines the growth of knowledge and skill, the application of machines and the availability of money and time. Knowledge and skill gained from education, training and experience become invested in people. Analysis involves the quality of the understanding held by all the parties involved and the effectiveness of the communication between them. This is influenced by the resources available at a given time.

When? Production equipment in the form of temporary works, and plant and machinery, allow for the development of solutions that break out of the limits of individual human strength. The production equipment must be conceived and constructed and hence forms a technological chain to provide the means to use processed materials and manufactured components, which can be assembled on-site. The building will then be used, maintained and adjusted to suit changing needs over its service life, which will require further use of machinery and resources. Eventually, plant and machinery will be required to disassemble the building and repurpose valuable resources into new uses.

How? Money and time, although not used directly in the technical solution, enable the available resources to be brought together so long as society, the client and those who seek to earn their living in the construction of buildings see the endeavour as good value. The realisation of the project requires management, organisation and control.

6 The Resource Base

This chapter introduces the key resources required for the successful creation and performance of buildings. Resources are normally associated with the activities of management, organisation and control. However, a knowledge and understanding of resources is central to the analysis of buildability, cost and social interactions in the evaluation of the suggested solution. The whole concept of buildability is centred on the capacity to successfully obtain resources to achieve the required performance criteria. Materials are identified as the primary resource, being associated directly with the forming of components and the construction of the final building. The second is the knowledge and skill held by people undertaking design and production to exploit the properties of the materials. Production equipment allows the application of materials in forms, sizes and shapes not possible with human effort alone. Money as a resource must be deployed by clients and the players in the construction industry. Costs arise only from the deployment and control of resources, the price of which are determined by the prevailing economic and political conditions. And decisions are taken in a timeframe dictated by demand. Society has an interest in the impacts on the physical and social environments of the use of resources for the process of construction, as well as the newly created resource of the building when in occupation.

6.1 Materials – the primary resource

Materials are a fundamental aspect of choice for the construction because of their properties, their behaviour to achieve performance, and their environmental footprint. There is a need to understand our ability to bring raw materials to a finished state in several component forms suitable for incorporation into a final assembly in a sustainable and ethical way. It is bringing the raw material to make components, followed by fixing in the final position, that determines all other resources required. The use of a material is, therefore, dependent not only on its potential properties, but also on the availability of all resources to process, manufacture and assemble the building. This needs to be seen in the life cycle of the building and the future reuse of materials once the building is decommissioned and disassembled.

In societies where resources are not readily available, construction is often limited to relatively unprocessed materials of local origin. The properties of the material may be difficult to predict due to variations in the quality of the material. This means either a process of

Barry's Construction Technology: Analysis and Choice, Third Edition. Stephen Emmitt and Tony Bryan.

selection giving a high level of waste (often economically unacceptable), or the design must assume only limited performance from the material, leading to larger sections of material to prevent failure. Many of the developments in the processing of materials seek to limit the variation in quality, as well as enhancing the basic natural properties to improve performance. The consequences of an increased manufacturing base and infrastructure are not only a guaranteed quality of materials, but also a greater range of materials. The existence of these new materials does not, however, lead directly to an ability to incorporate them into a design with the expectation of successful construction. There is always a need to be able to detail with the confidence that the required skill to work the materials for assembly into the building is available.

In addition to understanding the properties of a material for performance, its workability properties must be known. Very few materials are usable in the form in which they are found in nature. There are whole industries concerned with the processing of materials upon which buildings rely for the availability of components. In some cases, such as timber, this is mainly careful conversion, so that the material's properties are maintained. In other cases, such as plastics, formulation and processing can determine a wide range of properties, although only within the basic behaviour established ultimately by the long chain molecular structure of this group of materials. In both the cases of timber and plastics, not all types have similar workability. Straightness and closeness of grain for timber and the basic distinction between thermoplastic and thermosetting plastics, for instance, influence how they can be used.

The ease with which it is possible to perform operations on materials is key to choosing size and shape for components as well as possible fixing methods to provide the assembly. The ease with which each material can be cut, bent, poured, extruded, etc. without loss of its useful properties determines the manufacturing and assembly methods. This in turn determines the economics of the construction. This may lead to a range of materials being used for similar components. Bricks, although usually associated with clay that is moulded or extruded when wet and then fired, can also be made of concrete, which relies on chemical reactions that take place at ambient temperatures. Earth is another material that can be used. While this gives a product with many similar properties, the performance is different. Clay bricks expand in early life, while concrete bricks shrink; clay bricks can be dimensionally and visually irregular, while concrete ones have fewer deviations. Compacted earth bricks need to be protected from the elements to avoid erosion from moisture. These are endowed by the properties of the material; they are not a basic attribute of the component.

The materials we use are not limitless and their conversion and processing have an impact on both the built and the natural environments. Concern for the impact of extraction, depletion rates, energy for conversion, pollution, waste and ultimate disposal leads to the need for life cycle analysis. This indicates the true implications for the use of each material on the environment. It also concerns the ethical resourcing and processing or materials. Thus materials with all the right properties and that are workable with the skills available may not be chosen as a final choice because of their negative environmental impact and/or concerns about the ethics associated with a particular method of extraction and production. There are alternative choices available that claim to be more environmentally friendly compared to many of the traditional materials. In simple terms we need to consider reducing, reusing and recycling materials.

Reduce relates to reducing the amount of (virgin) material used and reducing the carbon footprint of the materials over their entire life. This will reduce the pressure on resource consumption, carbon emissions and construction waste (materials going to landfill).

Reuse is primarily concerned with the reuse of materials and components and viewing existing buildings as material banks. This requires consideration for the detailing, joining and fixing of materials to make it feasible to disassemble the building with minimal damage and waste. Reuse tends to use less energy and produce less carbon dioxide compared to recycling and processing materials into new uses. For example, reusing steel sections from a building that has been disassembled will be less energy intensive compared to melting down scrap steel and forming new steel sections. The challenge is to be able to source the components, in this example steel sections, in sufficient quantity and quality for the new project.

Recycling includes the recycling of construction materials and components, often to be repurposed and reused in new constructions and as new materials or components.

Management of (primary) resources is a fundamental part of circular thinking. The aim is to use nature-based materials, such as biobased materials, renewable materials and to a certain extent materials that are deemed to be ecologically efficient. Biobased materials include wood, bamboo, straw, hemp and cork. This includes biomass, which is a by-product of agricultural production, such as straw used for straw bales, and recovered and repurposed materials such as wine bottle corks turned into thermal and/or acoustic insulation panels for use in buildings. Materials that are already contained in buildings, such as steel beams, can be recovered and reused to contribute to a circular built environment. This feeds into a regenerative cycle with little or no waste of materials.

6.2 Buildings as material banks – the recycled resource

Existing buildings represent a considerable resource of valuable materials. These materials are locked into the building until such time as they are removed. Some materials and components will have a shorter service life compared to the building and will be replaced and subsequently recycled or reused. Once the building has served its function and is no longer deemed fit for occupation it will be disassembled and the materials and components will be recovered and reused. With a greater understanding of the potential and value of existing buildings as resources for future construction projects has come the concept of material passports. Material passports provide a digital record of the materials used in the building, their embodied carbon footprint and their health and safety properties. This provides essential information at a future date when the building is remodelled or disassembled following the principles of the circular economy.

Design for a circular economy

The most common design strategies that align well with a circular economy are design for disassembly (or deconstruction), design for adaptability, and design for optimisation. These are inter-related and overlapping principles that need to be considered within a reduce, reuse, recycle and regenerate design strategy. This will influence our analysis and choice.

Design for disassembly, also known as design for deconstructability or design for reversibility, is not a new concept. It has a long pedigree. The aim is to design and detail buildings in such a way that the components can be assembled and subsequently disassembled with minimal damage to the components. This thinking allows for individual components and larger assemblies to be demountable and easily replaced once they are no longer serviceable. This facilitates ease of maintenance, repair, recycling and reuse of valuable resources into new building projects.

Design for adaptability follows a similar philosophy. The aim is to provide the means of adapting a building by, for example, moving internal partition walls to suit a change in use or requirement. This can reduce the need for building work and may extend the useability and hence the lifespan of buildings with minimal waste or the need for additional resources. It requires careful design and detailing, sensitive choice of building components, and consideration for how building components are to be joined and subsequently re-joined in a different morphology at a future date.

Design for optimisation is a newer concept compared to the other two, although it does borrow heavily from well-known design principles of optimising materials, space and resources. One could argue these are the principles of good design and engineering and are easy to extend to circular thinking. Digital tools have made a significant impact on the ability of designers to explore many variables with a view to selecting the most efficient solution through optimisation. Optimisation software is a valuable tool in helping to reduce unnecessary use of materials and can assist with the desire to reduce waste. This does, however, sometimes lead to the questioning of existing building codes and design standards that include safety factors.

All three approaches should add to the reduction of embodied carbon and operational carbon if considered from the very early stages of the project. They will also contribute to the reduction of life cycle costs for building owners.

6.3 Knowledge and skill – the human resource

It is the development of knowledge and skill that determines not only the genesis of ideas about how we might build, but also our confidence to try. Human activity is predominantly social, and increased social complexity demands a greater range and purpose for its buildings. Given a sympathetic cultural environment, increased economic activity releases more resources to technological activity, which will be reflected in the methods used for building. Perhaps more significantly, the cultural environment determines the speed at which progress in technological development takes place. If developments move too far ahead of knowledge and skill, there is an increased risk of failure.

Every time a new idea is put into practice it is a trial that may fail due to an error in our understanding of the material or the behaviour of the construction. Technology derives a great deal of its knowledge and know-how from practice as well as from the sciences. It is an assessment of this total knowledge and skill that must be available before the risk inherent in the choice of construction can be evaluated. In a complex social environment this knowledge and skill is unlikely to be available in any one person or even a small group of people. Communication and collaboration between those with different aspects of technical understanding becomes increasingly important. Similarly, the ability for digital platforms to communicate seamlessly is essential in a digital built environment.

This resource, characterised as knowledge and skill, is identified in social organisations involving people. The operatives, designers, engineers and managers are most directly involved, gaining their knowledge from their education and training and then from their daily experience. Material and component manufacturers also generate and accumulate information and knowledge that is vital to the process of incorporation into the building. Behind all this lies the research and development activity often not directly related to the construction sector. This feeds in knowledge that can be called upon by practice to improve the chance of sound decisions and lower the risk of failure. This is enhanced by knowledge sharing and the use of automated digital knowledge management systems.

This knowledge is fundamentally, but not exclusively, technical. Design is a key ingredient in the choice of construction to provide fine buildings but is of limited value without an understanding of the technical issues involved. Organisational and motivational skills are vital to the manager but can be misplaced and ill-directed if the technical issues are not well understood. The skill of the operative to use tools cannot be isolated from an understanding of technical matters, particularly where new construction is being introduced. The judgements being made by the operative on, say, deviations (tolerances) are not just about the knowledge of the workability of their material but may have consequences for subsequent operations of assembly or the long-term performance of the building. It is the capacity of everyone to make technical judgements combined with their individual skill that is the key to identifying the availability of appropriate resources. It is the technical understanding that must be shared and communicated so that everyone may make sound decisions to limit the risk in the overall construction process.

6.4 Production equipment – the technological resource

With the right people with the knowledge and skill to select, process and work the materials into effective components and assemblies, it is possible to produce buildings. These buildings would, however, be made of a small number of materials and would be limited in size. While hand tools will be associated with the operative, it is the use of production equipment, both temporary works and plant and machinery, that expands the technological possibilities. More recently this has been coupled with digital technologies, which has further expanded the range of possibilities.

Although often seen today as the way to greater productivity and economy, production equipment was originally a resource that allowed solutions to be chosen that could only be built if the temporary works or machinery could be conceived and constructed. It formed a chain of technologies that had to be available before a particular construction could be realised. In the early days of large construction, such as cathedrals and early civil engineering works, these production technologies would have been designed and built by those responsible for the construction of the works themselves. They were often associated with lifting or simple materials processing. Manual skill would still be employed to shape and finish most of the components and fix them into their final position.

Although this production equipment is still fundamentally about breaking away from the limitations of the strength of an individual with only hand tools, the availability of these technologies may now make operations safe and economic and therefore become the preferred choice as they lower cost or increase quality. Knowledge of production equipment at the design and detailing stage becomes significant if safe and economic solutions are to be adopted. This process of mechanisation is not just associated with the site assembly process. Behind the expanding choice of materials lie developments in processing plant. Behind the expanding choice of components lie mechanised factory processes. The technological chain that allows choice of construction for the building is now extensive.

Plant and machinery have long been available for processing and lifting materials but often by mechanical advantage still using human power. Developments have been greatly influenced by alternative sources of a power supply. Individual power units can now be devised for very large and very small machines, making the availability and versatility of plant and equipment very great. However, most machines still require human knowledge and skill to guide and operate them. Perhaps the most radical change in the relationship between the choice of construction and machinery will come with the introduction of machines that take over the operator functions. With the application of microprocessor technology to automation of construction equipment, the conventions of the material–labour relationship may be supplanted by material/machine protocols. This can be seen in the use of computer numeric control (CNC), now available for structural steel fabrication, which can take its information directly from digital design information.

The use of power often involves the burning of fossil fuels, and this will have an environmental impact that needs to be considered in the choice of construction. Using energy from renewable sources, e.g. switching from diesel generators to ones powered by electric batteries charged by renewables is part of the transition to net zero.

6.5 Money – the enabling resource

Money is fundamentally a different resource in that it is not used directly in the construction of the building. It is not technical in nature and is therefore the most easily transferred to other areas of economic activity. Its use for building must be seen as good value by those who are willing to deploy it. Typically this is about the return on capital investment, and this determines much of the ethos in which construction is undertaken.

There are three major groups of individuals, or more typically businesses or corporations, for whom the deployment of money must be seen as good value. The first are the clients who commission the building. If the building can be built for the money the client is prepared to allocate to the construction, the client is likely to judge the building to be good value. The second group who require money as a resource are the professionals and the contractors who employ the client's money in the technical expertise and know-how of design and construction. Their attitude towards money, particularly the value of long- and short-term investment, is crucial to the technical potential of the whole industry. The third group is the building occupants and users. This disparate grouping of individuals may have no financial stake in the building; however, the decisions that have been taken will impact their perceived value of the building. There is another major player, the government. Grants and tax regimes influence the decisions of clients and investors in buildings. Similarly, and investment of taxpayers' money in public building projects fuels demand and workload.

Money is not a fundamental resource in the technical potential of a particular construction solution. This potential lies in the materials, the individuals with expertise, and the production equipment available. However, money plays a part in determining the levels at which these are available for construction and ultimately whether construction, however technically superior, will be applied and in what way.

Cost and value

The cost of a solution is important if for no other reason that ultimately a solution that is outside the economic reach of the client will not be built. It is also true that clients are unlikely to want to spend more than they need to obtain what they want. Two solutions of similar performance are likely to be chosen on price. What clients want and what they can afford may be two very different things, which may conflict. This will have to be resolved in the evaluation of a proposed solution. What clients want, and the proportion of their money they are prepared to spend on it, is related to the worth or value they place on the service they get from the building. This sense of value or worth that clients are likely to place on different features of the building must be established at the briefing stage, along with the more straightforward idea of the budget that will be available. Judgements on value will be made in different ways and these will influence the way in which cost should be analysed.

Cost and price

The cost incurred in adopting a particular solution for the construction of a building is computed from the price paid for the various resources employed for the goods and services to complete the work. Costs are created from a chain of prices charged which create the costs of each stage in the process. Hence the cost to the client for the construction of a wall is not the cost to the constructor. It is the price the client agrees to pay the constructor. The cost to the constructor is the price the constructor agrees to pay suppliers of goods and services needed to complete the works. If, therefore, it is required to analyse the cost of a solution, the question that must be asked first is: To whom? The usual answer to this would be the client, for it is the client that has the resources to invest, and the expectation of a return on the investment.

The scope of much analysis of cost that is carried out for the client is limited to the cost for the construction, the price paid to the constructor. This is often represented by the tender price. In the professional and industrial context there is considerable thought put into this concept of cost, for its commercial significance is high and hence data are available for its analysis. It is, therefore, tempting to use this as the basis on which a suggested solution for the construction should be evaluated for cost. This is, however, the client's cost, and it may not be the most appropriate basis on which to analyse the cost of a solution. The two major limitations in applying this basis of analysis are, first, that the cost data are historical, making interpretation of the figures for new or novel solutions difficult, and second, it does not consider the design or running costs, let alone the disposal and redevelopment implications.

Cost and time frames

The construction cost forms only part of the outlay the client will have to make. Even staying within the same time frame of cost incurred after land acquisition and prior to occupation of the building, there are costs associated with designing the works. Generally, where construction solutions are well established, the design costs will be lower than for novel designs. If experience and the gathering of existing knowledge still prove inadequate to determine the performance with an appropriate level of confidence, research and development costs may be incurred, although these are rare in construction.

The time frame of land acquisition to occupation also represents for the client a period of expenditure prior to gaining an income from the building. Designers and constructors will require payment in stages as the design and construction work proceed. The longer it takes to bring a building into occupation, the higher this cost. If the time between initial investment in the land and occupation can be influenced by the selection of a construction, it could be argued that this is a cost associated with that choice. Hence it could be seen as part of the analysis of its cost.

Different solutions have different costs after occupation of the building. Deterioration of materials suggested for alternative solutions will take place at different rates, requiring different renewal cycles to maintenance requirements. The reliability of components will vary, leading to different replacement costs. The detailing will determine the ease with which renewal and replacement can be achieved. This not only affects the direct cost of maintenance but also may carry a cost of disruption to the occupants while the work is carried out.

The overall building will have associated running costs directly affected by the construction choice. The example of the inclusion of insulation has already been mentioned and represents a part of the energy-use cost of a building. Another example is associated with the choice of surfaces, where the cleaning of a commercial building over its life may be many times that of the original construction costs. If cleaning could be achieved with less staff, albeit that the initial cost of the surface was more expensive, the overall cost might be less.

Finally, different solutions will have different disposal costs. This is not just the cost of demolition but also, increasingly, the cost of safe transport and deposit of waste from the demolition of the building. However, some materials that are waste to one project may have value to another. Recycling of architectural salvage has long had a commercial value, but increasingly building waste can also provide an income when the building is to be demolished.

There are cost implications associated with the land that must be purchased by the client. The availability of certain types of construction solutions will bring land into economic use. This is particularly true in the case of foundation and geotechnical ground improvement processes that can make poor land economic or allow usable space below ground. This can also be the case if lightweight superstructure solutions can lower the cost of foundation solutions.

Increasingly the development of brownfield sites where previous development has left land contaminated is becoming more viable as technical solutions to cleaning or capping sites are developed. This is also true if projected to the end of the life of the building to consider the redevelopment costs. While it is unlikely that new development will be allowed to contaminate land in the way past users have, certain foundation solutions such as piles may be difficult to remove. This will make the value of the land less for redevelopment at the end of the life of the building unless the piles can be reused to bear the new construction.

The analysis of the cost of a solution must be qualified by the question of not only whose costs are to be analysed but also over what time frame. It is unlikely that the full spectrum will be considered, as the client must balance the use of current capital against future income. A fuller cost evaluation using life cycle costing is necessary, and this will affect the analysis and choice of the construction.

Cost and cost data

Data concerning construction costs are readily available, however, they are predominantly based on historic prices for which constructors were willing to undertake the work once it was fully specified. The costs are based on measurements of parts of designed buildings and therefore rely on new work being like previous projects. There are reliable processes for estimating the cost of this measured work, but this is linked to construction costs of items measured in a standard way. Extending the requirements of a cost analysis beyond that of construction costs into extended time frames has the difficulty of establishing reliable cost data. While design costs can be estimated, it may be difficult to extract costs for designing a specific part of the building. Feedback on the cost of running a building may be limited, particularly for costs expected to arise some decades into the future. The lives of buildings are long and ownership often not continuous, so gathering these data is difficult. Even if data were available there would still be the need to project this into the future, where resource availability and economic conditions would be difficult to predict.

A more fundamental basis of analysis

Costs arise from the deployment of resources. The price that must be paid for resources depends on economic (and fiscal) conditions. For example, resources to be deployed for manufacture and assembly can be identified from an analysis of the operations required and the methods that could be used to complete those operations. It involves taking materials and components and operating on them with appropriate labour and equipment in a safe manner. This will give not only the quantity of resources required but also the time over which the labour and equipment will be employed. To these direct operational costs one must add any management and technical support and a proportion of the organisation costs, for these are also deploying resources.

In a market economy price is primarily influenced by supply and demand. When there is over-supply, the surplus usually results in price falls, whereas with scarcity comes price rises. In a competitive market the bid price will be based on a judgement of what others may be asking for the equivalent goods or services. The fiscal conditions influencing costs are either direct taxation or inducements such as employment schemes or regional development grants. The further into the future these predictions are made, the greater the chance of errors in the cost analysis. Running costs for the building in the first few years can be identified with some certainty, but the long-term renewal and replacement costs are often some years or decades ahead. Alternative resources may become available (or existing resources become unavailable), which may make an analysis of the expected operations and methods inappropriate. Changes in demand and supply or in the presence of new alternatives increasing competition will distort price differentials. New taxation will change prices, as happened in the UK with the introduction of VAT on maintenance work as well as new build. Future changes to regulations and building codes, for example relating to fire safety and/or health and wellbeing may make the predictions unreliable. However, decisions need to be made on the best possible information and analysis otherwise society cannot move quickly towards a more resilient and circular built environment. The predictions will not be perfect.

At the end of the building's life, disassembly may be made very expensive with the discovery of new hazards, as happened with asbestos, or made cheaper by the introduction of new recycling methods, as with concrete. All this means that estimating costs in extended time frames will be carried out with less confidence compared to the initial build costs. The figures will be subject to potential risks that are currently unforeseen.

Building up the price

It has been established that costs are built up from a cost–price chain where economic (and fiscal) factors influence price. Price can be fixed in one of two basic ways: what the market will stand or the application of rational margins. In practice there are likely to be aspects of both in the setting of a price, particularly in bidding in a competitive situation. In both cases, knowledge of the expected costs is important, plus some estimate of how certain those costs will be and the likely profit margin. The margin is the excess of price over cost that an organisation needs to survive and grow.

While full analysis of pricing is beyond the scope of this text, there is one more aspect that could be usefully highlighted at this stage. The relationship between cost and price should also reflect the risk. This may represent the technical risk of non-performance of the completed works and the subsequent costs this will incur. It may represent a financial risk associated with delay and non-performance of the contract. For these and other risks it may be possible to identify insurance that is at least a known cost that can be included. If, however, the risk is to be carried, some judgement on what price makes the risk worthwhile must be made. Generally it should be possible to get a lower price if you are willing to take the risk. Who in the cost–price chain takes what risk is dependent on the contract. Designs based on manufacturers' components and calculations transfer risk to the manufacturer. Developments such as serial tendering and partnering have the effect of limiting or even sharing the risk through shared understandings and information between the contracting parties. This is likely to affect the construction solutions seeking lower costs for all parties involved.

6.6 Time – the temporal resource

Time is also fundamentally different from the other resources in that it is a temporal resource. In the context of a construction project, time is a measurable parameter, similar to cost and quality. The timeframe for a construction project will be set out by the client at the briefing stage and this will determine the amount of time for each stage in the construction project and hence the amount of time available for analysis and choice. Time is required to make suggestions, analyse those suggestions and make informed choices. When dealing with familiar building forms, materials and technologies, the tried and tested solutions will be used and the amount of time required to make informed decisions may be less compared to that required to explore unfamiliar materials and solutions. And, as noted in the previous section, time is closely related to cost. It is important to allow sufficient time to make informed decisions during the design of the building, and also during the construction phase when it may be necessary to suggest changes to specified materials and components. Change management protocols will specify procedures for considering any changes to the design, for analysing the proposed changes within a reasonable timeframe, and proposing a solution.

Much of the attention on construction projects is the timeframe for the design and construction process to be completed, and this falls under the remit of the project manager. This is, however, only a small percentage of the time that the building will be in use. The time in use is significant and it is during this period that the building will be subject to weathering and wear and tear through general use. Viewing the building over its entire service life will help to identify the frequency of cleaning, maintenance, routine replacement and anticipated repair of failed elements. This is linked to life cycle costing and life cycle analysis. It is also linked to visualising the building as part of a closed economic loop.

Sustainability – Social Concerns and Technical Interventions: Chapter 7 AT A GLANCE

What? There are two major forces that condition a construction solution: the laws of nature and society. At any time in a society there may be deep changes in its values and beliefs, which must be considered in a technical solution for it to be seen as a reasonable option. This includes embracing environmental concerns in our understanding of sustainable development and regenerative design. In developing a response to the climate emergency the key areas to consider are reducing energy use, pollution and resource depletion. This can be tackled by adopting ethical approaches to construction, applying circular economy principles to analysis and choice, and using a variety of frameworks to inform our decision making to realise a resilient built environment.

Why? Sustainability is not exclusively concerned with buildings. However, as polluters and resource consumers in both the construction and operation of buildings, choices about how we build must be integrated into society's responses to the climate emergency. The collective decisions made about how we build will have a significant impact on the environment and on our social and economic wellbeing. These fundamental shifts in values often demand changes for which knowledge is uncertain, making the analysis of physical behaviour of the construction even more significant to ensure environmentally safe and durable solutions.

When? These concerns need to be considered at the very early stages in the design, as internal layout, size, space and orientation are important to principles such as passive design and reducing embodied and operational energy. The choice of equipment and building management also contribute to reducing environmental impact, for example through the specification of micro-generation of power with active building designs. When making the choice of construction, the detailing and the specification of materials becomes crucial to the building performance. This needs to be considered at the assembly stage, and any proposed changes to the design will need to be analysed against environmental and performance criteria for the entire building before making changes.

How? Sustainable and regenerative design principles can make a significant contribution to achieving buildings with low-energy demands, particularly in meeting the required thermal, ventilation and lighting performance. Similarly, the use of building assessment systems such as BREEAM (Building Research Establishment Environmental Assessment System) and LEED (Leadership in Energy and Environmental Design), in conjunction with computational analysis of performance, will bring about buildings that have positive impacts on the planet and the occupants' health and wellbeing.

7 Sustainability – Social Concerns and Technical Interventions

This chapter considers sustainability as a social concern requiring a technical response to address the climate emergency. Specific technical knowledge concerning sustainable development has been provided in other chapters, where it needs to be embedded in the process of analysis and choice. Sustainability is set in the context of social demands, requiring technical solutions that may sit outside the current knowledge base as embedded in building information models (BIM) libraries and familiar approaches to construction. Challenging the convention and pushing the boundaries of what may be possible requires a balanced approach to perceived risk and anticipated reward. This requires additional care in analysis and choice to weigh performance criteria against long-term environmental impact until such time as the knowledge base has been established.

7.1 Sustainability – two major forces

There are two major determinants of the construction solutions that we adopt. One is the law of nature, and the other is the social context within which we operate. These two forces condition the technology we develop and apply. This is the context in which we design, build and use buildings. We strive to understand the natural resources that provide our materials and the forces of nature that act on the final construction. This is the knowledge we use in the analysis of the physical behaviour. Our knowledge in this area is now well established, and while it will continue to improve it is unlikely to change fundamentally. We need to also understand the cultural, economic, legislative and political conditions that exist, and recognise that these will change with time. These factors shape building designs, determine the resources that are available for construction, and affect the commercial and ethical limits that govern the organisation of construction and the building in occupation. This influences the skills available and methods of operation that make solutions buildable, and which become the major determinant of the cost of construction. Changes may require new technical knowledge, especially with the need to return to our fundamental understanding of construction systems and systems behaviour to address the climate emergency.

Any extended discussion on the nature and impact of cultural beliefs and values and an ethical framework is beyond the scope of this text. However, when society is questioning its own values and ethics it is inevitable that these will be reflected in the questioning that has to be undertaken about what can be 'reasonably' achieved and what is a 'reasonable' solution to the challenge. This leads to changes that have technical consequences. Technology

Barry's Construction Technology: Analysis and Choice, Third Edition. Stephen Emmitt and Tony Bryan.

must respond to shifts in values to be accepted by society in a continual process of evolution and change. As those involved in construction are usually part of the society they serve, or are at least culturally linked to that society, most of these influences are inherent in their collective decision making. The social concern for our impact on the natural environment is having a profound effect on the design of buildings, the way that we build, and the way in which we use and subsequently recycle buildings. It is making us question the available technologies and extant knowledge in our process of suggestion, analysis and choice.

Sustainable development and regenerative design

All generations have worked to improve the quality of life for their children. As societies develop, this has been dependent on an economic process, the creation of wealth, and then in its distribution that determines social equity, providing the rewards of economic development to individuals. Linked to political beliefs associated with values and ethics and the best ways to stimulate economic growth and social equity, this has determined the demand for buildings and provided the means to produce them. In recent years we have made such progress in economic and social terms that development is now at an unprecedented scale. This has brought two great forces together, the natural environment and society. Development is now changing the environment to such an extent that we must face the possibility that we will not have the resources to maintain our current rate of development or, even worse, may make the environment so hostile that we will not be able to provide technological solutions to maintain and improve the quality of life. It is now widely believed that the choices available and the quality of life for the next generation will be harmed if we do not bring together concerns for the natural environment as well as our concerns to maintain stable economic and social conditions. Fears for the natural environment must inform our analysis and be a central part of our process of choice to decarbonise the built environment. In the same way as the choice of construction is mediated by economic and social conditions, so it will need to consider the building's impact on the environment.

Many of the economic and social influences seen as important in sustainability are well established and have been shaping our designs and the construction solution for many years. What has changed is the urgency and the scale of change required to address the climate emergency. This has generated the need to identify the important issues and look for knowledge and understanding that is sufficient to suggest novel, yet safe and resilient solutions. Successful construction will require the existence of complementary knowledge and the know-how of all involved. It becomes part of the analysis of a solution to ensure that this complementary expertise exists before choice is made if performance is to be realised. This highlights the need for education, training and research and development to be a part of society, and of the industry and professions that contribute to construction.

The concept of sustainable development introduces environmental impact into decisions on construction detailing and specification. The analysis brings into question the origin and ethical sourcing of materials, their manufacture, assembly, disassembly, recovery and reuse; as well as their contribution to pollution, energy use, resource depletion and changes in ecosystems. A recent development has been to see the entire construction and use of the building as a circular (or closed) system. Another development has been to see sustainability a sub-set of a wider regeneration of our built and natural environment. The concept of regenerative design is founded on the principle of human and natural systems co-existing and co-evolving over time. The value of this approach is that it has the potential to regenerate planetary health and have positive benefits for people and the environment. While sustainability is primarily concerned with reducing and limiting resource use, regeneration aims to replenish resources and natural systems. Thus regenerative building design should be a fundamental driver of analysis and choice.

7.2 Responding to the climate emergency

It is important that in providing a response to the climate emergency that we are asking the right questions. Sustainability and regenerative design are wide concepts, exercising the thoughts of many sectors of society. Even the simplest analysis shows that construction and buildings in occupation have a significant environmental impact. It is also clear that, while not directly associated with the technological solutions adopted for the buildings, development and urban design has implications for travel and journey times that will influence the transport sector, which also has a high impact on the environment. It is the nature of these problems, their interactions and interdependencies, that play a significant part in the analysis. The concerns of this text are with construction and buildings in occupation; however, it is necessary to be mindful of the environmental challenges for all aspects of human activity so the contribution made by construction can be effective. The concerns include global warming and climate change, ecosystems and biodiversity, resources and waste management. There are two distinct issues that can be identified as common threads, pollution and resource depletion.

Pollution has impact associated with changes to the environment. Its effect may be direct in poisoning the air, water, or land and thus influencing, among other things, ecosystems, biodiversity, and the food chain. It also has indirect effects. One of the major concerns is global warming. Burning of fossil fuels releases carbon dioxide that we now recognise as a greenhouse gas. When it accumulates in the atmosphere, it permits the short-wave radiation from the sun to heat the earth but limits the long-wave radiation from the warmed earth back into space, resulting in warming that is changing our climate at a rate that life on earth may find difficult to respond to. Thus we characterise CO_2 as pollution, although there are other greenhouse gases such as HCFCs (hydrochlorofluorocacarbons) and nitrous oxide (NOX) that are produced in the operation of the building.

Resource depletion affects the continued availability of the resources with which we construct and maintain our quality of life. Some resources such as timber and water are identified as renewable, but this must be managed as demand can still outstrip supply. They are part of natural cycles that, if not respected, will create conditions that will make supporting our aspirations for quality of life unobtainable. Some resources are acknowledged as finite (either physically or economically) and some of these could run out (or be priced out) if we continue to use them at the current rate. Reducing the availability of resources and maintaining a continuity of supply may have a significant impact on economic and social systems before they have irrevocable environmental impact. Building has always had an impact on the environment and resource availability. In the past the impact has often been local, and it has been within the capacity of the environment to recover with time. When we judge recovery, we are generally using the criterion that change does not significantly threaten human life or the quality of life both now and in the future. We are now facing the probability that recovery is not possible and that we may not have the technological capacity to deal with the range of conditions to maintain the wellbeing of the human population. We must reduce and reverse the adverse impact we are currently having on the environment.

It is now recognised that we must look for a convergent solution. For example, it is generally accepted that current levels of our use of energy and resources cannot be replaced by managed renewable sources in the timescale required. We need to reduce demand for energy while increasing the use of renewable sources; but, for the near future, we will still have some need for fossil fuels as we shift to electricity generated from renewables. It is one of the significant areas of analysis if buildings are going to reduce their impact on the environment, but it is not the only concern. There is a need for consumers to rethink their behaviour. This is perhaps most evident in the mantra to reduce, reuse, recycle, revitalise and regenerate. Building users are also consumers of materials, water and energy, and produce waste. Buildings must be designed to have a low environmental impact in operation and low environmental impact in construction, maintenance and disassembly. The two must be considered together. For example, concrete can be characterised as being environmentally unfriendly because of the energy used to produce the cement. However, if the cement is sourced from factories using renewable energy, specified with cement replacements, used in the design for its thermal mass to save energy throughout the life of the building, and detailed to be disassembled and reused or recycled at a future date, it may provide well provide a solution that is not as environmentally unfriendly as it may seem at first glance. A similar argument could be made for steel and timber. The choice of materials has become a complex undertaking that now relies on lifecycle analysis (LCA) and life cycle costing (LCC) to guide our decisions.

Limiting environmental impact

It is necessary to see the building as polluter and resource consumer as well as contributor to economic and social activity. Like the building's economic and social contribution, it is also necessary to see the building's environmental impact through its whole life cycle: construction, use, disposal and reuse and redevelopment. The process of analysis and choice already takes many factors into account, from physical behaviour and appearance to production and cost considerations. Looking for low environmental impact is just one more factor in this process. By considering the creation of the internal environment through heating, lighting and ventilation, low-energy solutions can be identified. Looking at processes such as water consumption and waste systems, the impact on the environment can be reduced. Making choices about components and materials while considering their origins, embodied energy and waste implications leads to less environmental impact.

It becomes helpful to have a framework into which the choices about building design and detailing can fit into the concerns of sustainability. The United Nations (UN) have set out sustainable development goals (SDGs), which can guide clients and their professional advisors to create and maintain better buildings. From these a series of goals and strategies are set that can be adopted by communities. Some of the areas identifying action are more applicable to building and building development than others, but it is possible to see connections to most and therefore to identify actions that can be taken throughout the cycle of building development. These will then influence the specific activity of selecting technologies and preparing specifications and details. The emphasis in regenerative design is on communities, so it becomes necessary to ask how buildings contribute to communities. It is then possible to apply the principles by adopting appropriate strategies in the design and choice of technologies employed in creating, operating and ultimately disposing of the building.

Frameworks suggest that there are many ways in which a building can contribute to reducing its environmental impact and support sustainable development. With the introduction of regulations and governmental pledges to meet targets comes the need to measure a building's contribution to sustainability. This has resulted in the development of environmental assessment rating systems, such as BREEAM and LEED. These are essential tools for guiding analysis and choice.

7.3 Buildings as systems

Visualising buildings as a system of a series of flows and transfers is fundamental to understanding the dynamic behaviour of buildings. This again proves useful in evaluating building design against environmental impact criteria. By looking to minimise inputs, make the best use of resources while they are in the building, and then minimise outputs, many of the aims of low environmental impact can be achieved. This can perhaps be best seen in what has become known as passive design. If flows of heat, light and clean air can be kept in balance to maintain the required internal comfort conditions through the passive (and active) fabric, the building will have a low environmental impact. Services use energy and therefore designs that can maintain the balance of flows passively make a significant contribution to reducing energy use and carbon reduction. The main users of energy are heating, cooling, lighting and ventilation, so these become the focus of passive design strategies and low carbon technologies.

Passive environmental design engages not only the fabric for gains and losses but uses the spaces in the building to introduce natural light and capture heat that can be moved around the building by natural forces. The aim should be to minimise the need for active systems that use energy. Active systems that are still needed can then have efficient components and be controlled via building management systems (BMS) and the energy still required then can be supplied from renewable energy sources to further reduce carbon emissions. These three aspects of passive design, energy management and renewable generation all need to be incorporated at the concept design stage. It is less effective to add them later in the design of a new building. However, efforts to improve the energy efficiency of existing buildings can make a significant contribution to reducing energy use and CO_2 nationally.

Passive design for thermal environments

In the UK, the seasonal variation is such that there is almost certainly a need to heat in the winter months and cool in the summer. This will be influenced by internal heat gains that will contribute to heating in the winter but will increase the need for cooling in the summer. In practice this often means that for housing the major concern is winter heating, but for commercial and industrial buildings there is a significant number of days in the summer when cooling will be required. It is therefore necessary to have passive designs that can perform during both winter and summer conditions. Although this analysis is for comfortable

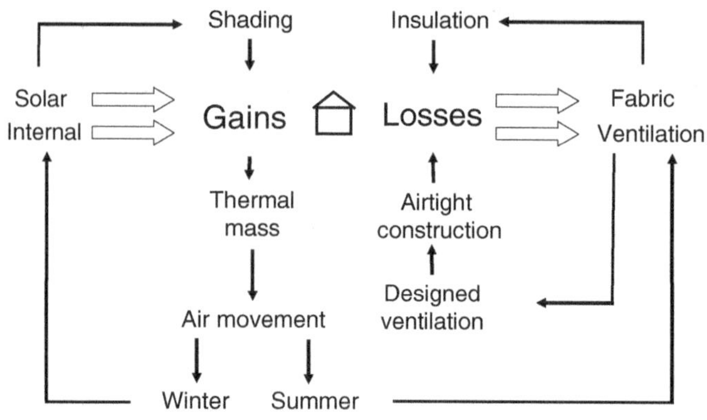

Figure 7.1 Thermal gains and losses.

thermal conditions in the building, it will be seen that the requirement for ventilation for clean air must also be considered and will become part of the design requirements.

Figure 7.1 shows the flows that need to be considered to maintain the thermal environment by focusing on the analysis of the gains and losses from the spaces in the building. The design needs to ensure that there is a balance at the desired internal conditions. Heat gains come from solar gain and internal sources such as the occupants and the operational equipment as well as building services such as lighting (although a successful passive design would minimise gains from these services). These are useful in the winter but unhelpful in the summer. Passive losses occur through the fabric of the building and ventilation. Fabric losses can be limited by insulation, and this makes a significant contribution to limiting losses under winter conditions so the need for high levels of insulation is now well accepted for all buildings.

Ventilation is necessary to provide clean and wholesome air and to control humidity levels. This can account for heat losses in winter and therefore ventilation needs to be limited to that required for healthy living. In summer ventilation is used to help provide the correct thermal environment with the introduction of cooler air and increased air speed to help evaporation to cool skin. To satisfy the needs of both summer and winter thermal conditions and wholesome air environments, there is a need to build airtight buildings and then design sufficient and controllable ventilation to provide good air in winter and direct cooling air in summer. In passive design this ventilation should be achieved using natural forces provided by the wind and thermal differences within the building. Rooms may be ventilated naturally by single-side ventilation, but if the internal layout and building shape and dimensions allow then cross ventilation and the stack effect can be utilised, as indicated in Figure 7.2. These will be considered in more detail below.

There are relatively few days in the year when there is a balance between these passive gains and losses to give a satisfactory thermal environment in all rooms. In the recent past, active services would have been designed to bring the conditions back within desirable limits. However, two other passive mechanisms, thermal mass and shading, can be brought into play to hold the balance between gains and losses.

The first mechanism is to exploit the thermal mass of the fabric. The fabric is used to store heat during times of high gains and release it when losses are possible. This proves useful

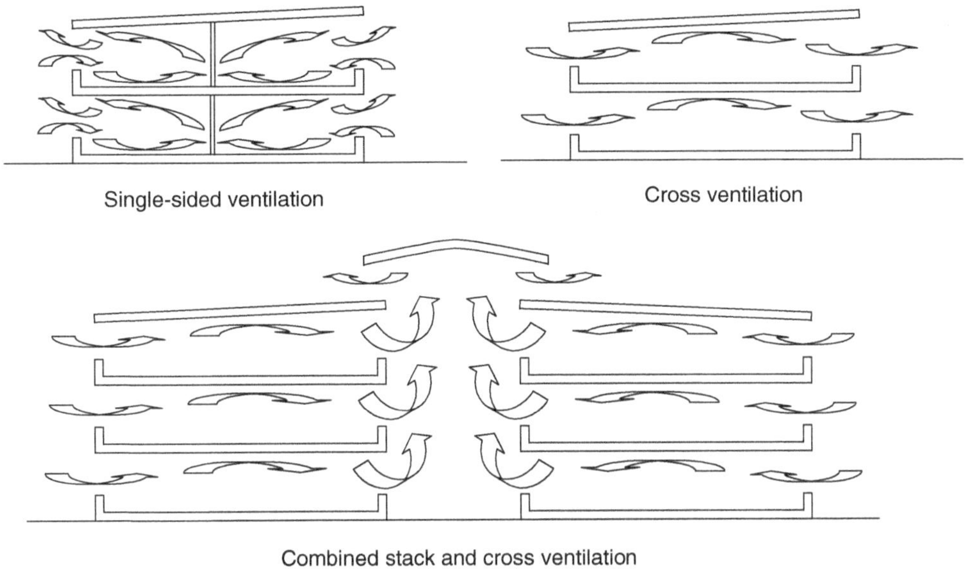

Figure 7.2 Side, cross and stack ventilation.

over the 24-hour cycle where daytime gains are high and night-time losses can redress the balance. This time shifting of gains and losses must be controlled and will require the heat to be moved around the building. The most effective passive mechanism to move heat is to use the air in the building. This requires predictable and controlled air movement through the building. This may be different from the ventilation process associated with clean and tempered air. It is necessary to have different air movement patterns for summer and winter.

In summer, the movement is to provide ventilation as a loss during the night, known as night purging; while in winter the heat must be treated as an internal gain with the heat being distributed inside the building, while ventilation is running at a minimum for wholesome air. The forces that move the air in summer are likely to be those being used for daytime ventilation. Single-sided, crossflow, and stack effects can be used, but the air must be directed towards the thermal mass being used for storage not over the occupants to increase cooling. The position of the thermal mass is important for both effective storage and transfer during purging. The most usual part of the structure is the underside of the floor slab. This approach is shown in Figure 7.3. Hot air rises in the day and the thermal mass in the slab will start to absorb the heat. In addition to absorbing the heat, the high thermal capacity will mean that the soffit of the slab will take time to rise in temperature. All the time the surface temperature is below air temperature, the occupants will experience cooling as they radiate heat to the slab. When the air outside cools sufficiently to receive the warm air in the building, windows must open automatically to start the night purge, drawing in the cool night air that passes over the slab soffit and then ventilating back out into the night air. This 24-hour cycle utilises about the first 40 mm of the slab as effective thermal mass. However, the whole thermal capacity of the slab is utilised indirectly as any thermal mass will absorb heat during the day and will give it up to cooler air at night. Thermal mass will assist the gains–losses balance, and this is particularly useful under winter conditions.

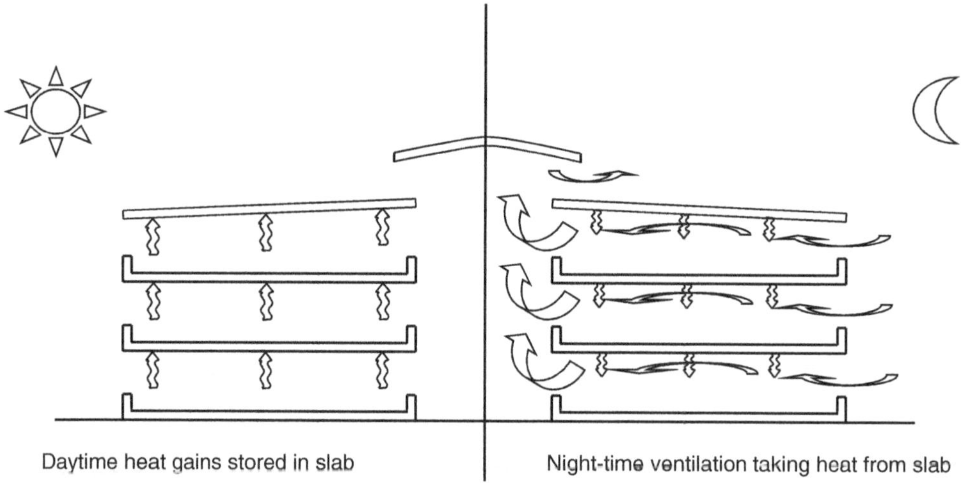

Figure 7.3 Summer night purging (commercial building).

In winter, the night-time air movement must remain inside the building and so cannot take advantage of the external pressure driving the crossflow and stack effects. The only forces come from thermal gradients and perhaps the small amount of ventilation for wholesome air. The other difference between summer and winter is that in summer it is unwanted heat in the air which is taken into the thermal mass while in winter it is solar and internal gains that are to be captured. This heat can then be used to limit the fall of the internal air temperature at night when no significant gains are available.

Daytime internal gains will be collected by any thermal mass and redistributed during the night, even if only back into the room in which it was collected. Rooms that continue to have internal gains through the night may need some provision to vent the air from the room to other parts of the building. For the maximum effect from solar gains in winter, the thermal mass must be internal and in direct line with the sun. This means facing south and behind glass. This condition will create significant summer difficulties in most commercial buildings but can be useful for house design. Both heavy walls and floors can be used to capture heat from conservatories, windows and roof glazing and help keep the house warm during winter nights when losses are at their greatest. This approach is shown in Figure 7.4. There is now a potential for rooms such as conservatories to overheat in the summer when it may be difficult to keep this heat from being transferred into the house during the day.

The dilemma between using solar gains for winter conditions and the overheating problems this may cause in the summer leads to the second passive mechanism: shading. Solar gains are greatest from the south but even walls and roofs orientated more to the east and west will have direct solar gains for part of the day. While it is possible to use internal shading such as blinds, this has limited effect as they will be warmed by the sun and this heat will then be inside the building and become a contributing gain. External shading in the form of shutters, brise soleils, secondary louvre walls, or natural planting is more effective. Brise soleils can be arranged to cut off the sun when high in the sky in summer but allow the low winter sun to enter to make the most of potential gains. This approach is shown in Figure 7.5. Deciduous trees can similarly shade in summer when in full leaf but let through winter sun when the leaves have been lost.

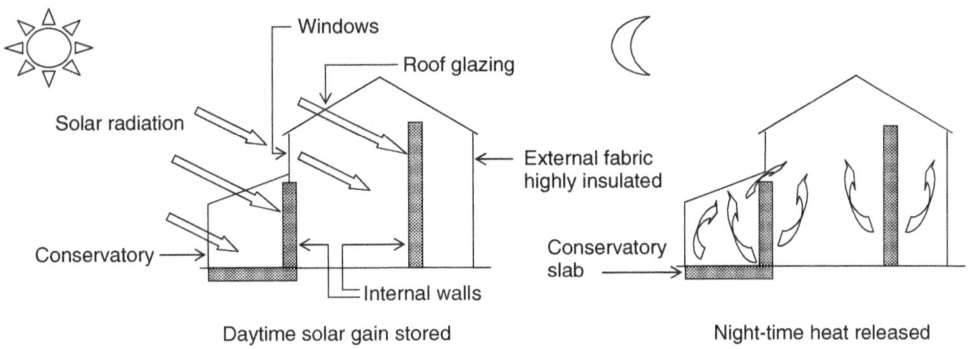

Figure 7.4 Winter heat gains (domestic building).

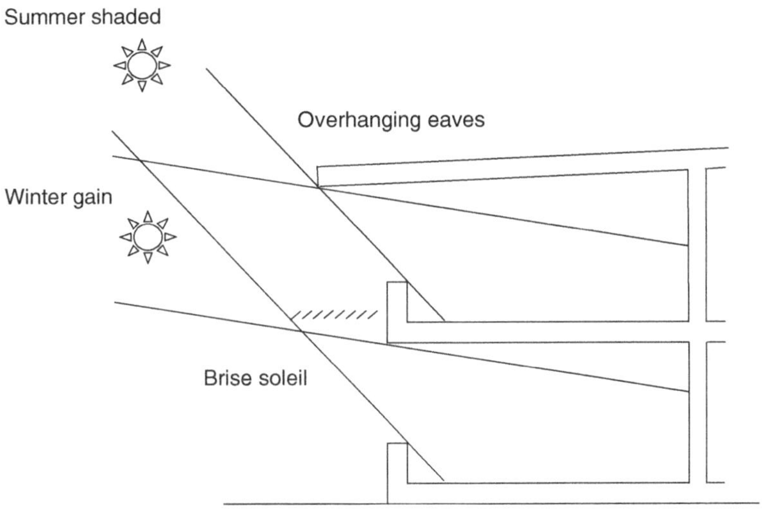

Figure 7.5 Winter gains, summer shading.

Building form and internal layout

If flows and transfers are to be exploited, the need to use natural ventilation will have a significant influence on building form and layout. As natural ventilation is using only small forces to move the air, the size and proportions of the spaces and sizing of the openings need careful attention to ensure that sufficient air flow is achieved. The design can be modelled using software, with predicted airflow modelled for a variety of conditions using computational fluid dynamics (CFD). In simple terms, the design needs to consider the positioning and sizing of the vents. Pressure differences are created by wind flowing around the building giving positive and negative pressures that can drive either single-sided or cross ventilation. If cross ventilation is to be employed, the width should not exceed around 4.5 times the floor-to-ceiling height. For single-sided ventilation the depth of the room from the windows cannot be more than around 2.5 times the floor-to-ceiling height.

Floor-to-ceiling heights are unlikely to exceed 3 metres, giving a limit to the building width of around 13.5 metres for open-plan layouts with cross ventilation and perhaps 16 metres if there is a central corridor for single-sided ventilation. Both cross and single-sided ventilation need sufficient areas of opening window, and for single-sided ventilation openings at low level for inlet and high level for exhaust can enhance ventilation rates with a local stack effect.

The stack effect is a vertical movement of the air that can also be used to move air through the building from openings on each floor level, normally to openings at or above the roof level. The stack can be driven by buoyant warm air rising through the building and out at the top. This can work even in the absence of wind on still summer days. However, in peak summer conditions, if the passive design works well and shading and thermal mass maintain low internal temperatures, the stack may not work well during the day until the building has started to overheat. Buoyant thermal effects will work well at night for effective night purging.

The stack effect is also driven by the wind-induced pressure differences that create the cross and single-sided ventilation. On all but still days this is likely to be the predominant force moving the air in the stack, particularly if used in conjunction with cross ventilation. The top of the stack must be high enough to ensure a lower pressure at the top of the stack than at the window level on the highest ventilated floor or a reverse flow will occur at this floor level. It is also possible to use the wind direction to introduce fresh air at the top of the stack. This requires some mechanism to ensure that the inlet section of the vent is facing into the wind. The stack can be created from vertical open spaces such as atria. If stack shafts are to be used, there needs to be around 1 m^2 of area for each 6-metre section of building to serve three storeys. Smaller stacks can be used but they may need fan assistance.

Natural lighting as a passive system

Artificial lighting uses energy, and therefore designs that maximise the use of natural light can make a significant contribution to passive design. The analysis of passive thermal design utilising natural ventilation led to narrow plan buildings that also give benefits in using natural light, as do atria used for stack ventilation. To give required light levels towards the centre of even these narrow plan buildings there will need to be more glazing than is required for the openings necessary to ventilate. The penetration of light into a room is dependent on the height of the head of the window and the width of the room. It is also dependent on the reflective values of the surfaces in the back half of the room. For example, a room 3 metres wide with a window head height of around 2.5 metres and a surface reflective index of 0.5 will only be possible to light naturally up to around 5.5 metres from the window, however big the window. However, a room with the same window head height but 10 metres wide with surface reflectance of 0.6 could be lit to a depth of around 7.5 metres, matching the room depth for maximum natural ventilation. These room dimensions only show a potential for natural lighting. The area of glazing must be designed to achieve this potential although this may aggravate unwanted thermal gains. As with thermal gains, orientation is a significant factor in lighting. North light is consistent, while light from the south has the potential for high light levels that can produce glare. Light from the south may need shading, particularly in summer. External brise soleil used to limit thermal gain would provide shading in the summer but not winter when internal blinds may still be necessary to control glare.

Buildings clustered around atria or courtyards for natural ventilation can also use the atria or courtyard to bring light into the building if they are wide enough to ensure adequate light levels in the absence of a direct view of the sky. Although natural lighting must be considered separately to the thermal design and may need more glazing and/or may limit building widths, the general building form and orientation for both thermal and lighting requirements are compatible. These design considerations for both ventilation and lighting make deep plan building difficult to design passively unless the internal layout can create paths for the air movement for natural ventilation and lighting. However, narrow plan buildings can be grouped to provide streets or courtyards with either covered atria or open spaces that allow for façade ventilation and lighting to both sides of the internal spaces. These arrangements are shown in Figure 7.6.

Mixed systems

Even with passive design it may be necessary to provide active systems to deal with conditions towards the extremes of the design conditions to keep the gains and losses in balance. This may be particularly true on difficult sites where space is limited and there is little choice in the configuration and internal layout of the building. Some form of occasional heating will probably be required in the winter and energy may be required to ensure hot water all year round. With high internal heat gains, it may be necessary to have fan-assisted stack ventilation or even some areas mechanically ventilated or cooled to increase losses in summer. Heat recovery from exhaust ventilation air may prove to be energy efficient if it can recover more energy than it consumes to run. Areas that are difficult to adequately ventilate naturally, such as kitchens, bathrooms and toilets, may require mechanical ventilation to remove odours and humid air and replace it with wholesome air. Low-energy artificial lighting will be required for night-time operations and to supplement daytime lighting levels.

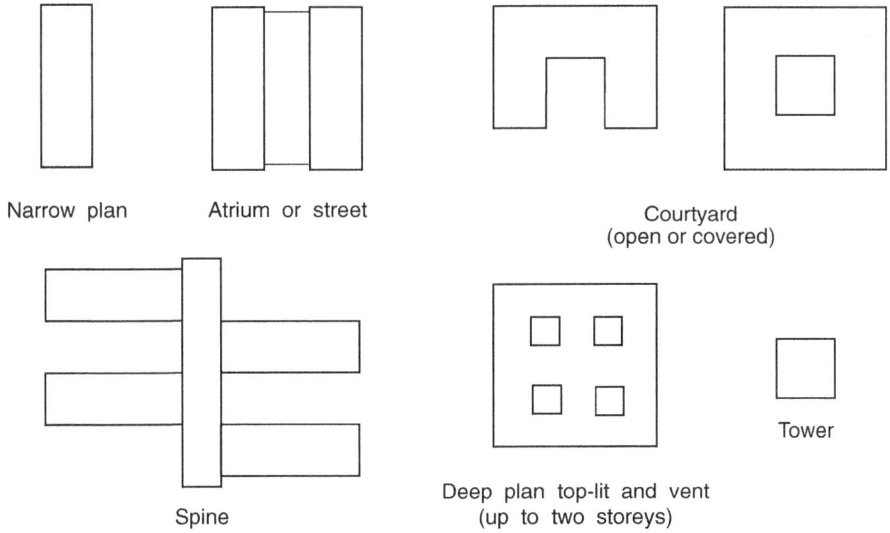

Figure 7.6 Building layouts capable of using natural ventilation and lighting.

Reducing carbon

Passive systems aim to create and maintain environmental conditions using natural forces, although some use of active fabric will need small amounts of acquired energy. The services systems plus the operational equipment such as computers will need energy. There will be equipment specified within the building that will need energy. Energy-efficient versions should be specified, such as LED lighting and air source heat pumps.

Beyond passive design it is possible to reduce carbon emissions of active systems by careful choice of equipment and management of the systems using data from the building in occupation. BMS can make a significant impact on the energy use of a building and assist with the management of spaces. Full BMS are linked through digital internet-based systems and normally associated with commercial buildings, although such systems for homes are available. The management part of the system is about control. It makes use of remote sensors to collect and provide data that informs the system. As was indicated above, automatic control will be necessary if natural ventilation and night purging are employed in passive systems with active fabric. The opening of windows will need to be on a temperature and/or time base to ensure the air flows at appropriate times. Motorised vertical shading that follows the sun would also need automatic control.

BMS can also be used to control the active services to reduce energy consumption. The simplest sensor systems do not need computer control and can be directly linked to switching the energy using appliances. Thermostats can be used to monitor temperature and turn heating systems on and off while movement, sound or light sensors can be used to switch lights off and on. Movement sensors detect an absence of people and therefore turn lights off (and back on if movement is detected), while light sensors, often in the light fittings themselves, detect increasing levels of light and turn off any lights that are switched on. The most effective use of light sensors is to have lights wired in banks parallel to the windows so changes in natural lighting levels that normally occur across the depth of the room, away from the windows, can be compensated for with the artificial lighting. This will maintain lighting levels without the need for the occupants to respond to the changes in light. The occupants are less likely to switch lights off as natural lighting increases, although they will switch lights on as lighting levels drop. None of these automatic controls remove the need for switches that the occupants can use to override the automatic systems. BMS can also incorporate the security and fire detection systems that also use sensors to trigger alarms and sprinkler systems.

The information function of the BMS should be designed to give the user and building managers instant and time-based information for both heat and power usage. This should be related to specific areas of the building where occupant behaviour can be investigated and informed in a way that will modify behaviour to reduce energy consumption. This may require the building operator to inform and educate the occupants in how best to use the systems to maximise energy efficiency. This may include everyday behaviour such as switching off equipment overnight but may also draw attention to the way the building is being used. Energy-efficient appliances and simple sensor systems such as thermostats and light sensors can be installed without computer control, but there is then a need to inform the users of how to make the best use of these systems. Here the information should be available through a user guide. This should inform the occupants of how the systems should be operated and used to their full advantage. User guides are particularly significant in homes where there is unlikely to be a building manager to inform and train staff. The occupants need the information specific to the installation in the dwelling and need to take full responsibility for maximising energy efficiency. Smart meters will help provide this information.

7.4 Renewable energy sources

Energy will be needed in most buildings. In seeking the convergent solution to energy needs the use of passive design and system efficiency reduce energy demands but there is still an opportunity to reduce CO_2 by considering supply from low- or zero-carbon technologies. While this policy of renewable energy sources can be pursued at the national level, it is also possible to look to technologies for individual buildings or small community schemes. These are often known as micro-generation technologies that employ photovoltaic (PV) panels and small wind turbines to generate electricity. The term active building is used for buildings that generate more energy than they need to operate. These systems also become more viable if the electricity can be stored in batteries or fed back into the national grid when generation exceeds the demand from the building. Wind for the individual building, particularly in urban areas, is unlikely to be efficient unless the development is large enough to install a large independent mast-mounted turbine. Proximity to a fast-running stream offers the possibility for hydroelectricity, but this is relatively rare on development sites. Both wind and water will be used for national generation schemes for distribution via the grid.

Beyond the carbon-free sources there are systems that can be considered carbon-neutral. These generate heat and power from plant-based biofuels that can be considered carbon-neutral in so far as the carbon that is released is captured in the next generation of fuel plants currently being grown. For these to be truly neutral and renewable there must be a growing programme of supply to match demand. Unfortunately, this is limited as the plant-based biofuels are competing for land for food growth and natural habitats for biodiversity. Leaving this to the market to seek a balance on price may not achieve the sustainability goal of limiting environmental impact or social equity.

Beyond carbon-free and carbon-neutral, carbon-efficient systems such as heat pumps can be used to draw heat from the air or the ground to provide the space heating and possible hot water for the building. These systems do use power to transfer the heat, but the heat comes from the sun that warms the air and the ground and is, therefore, carbon-free. Even air at below 0 °C has heat in it, so air systems work in winter conditions. Air pump systems normally move the heat from outside air directly to inside air so either are room-based systems or need internal air movement if units are to heat more than one space, clearly best supplied by natural air movements within the building. Ground pump systems draw heat from depths where the ground is naturally less variable in temperature than the air. These need not be too deep as surface systems laid in trenches around 1-metre deep can be effective but do need a loop of around 200 metres for an average house. Where this area of ground is not available, systems can be installed in vertical bore holes.

Ground source heat pumps are suited to exchanging heat to the water of a central heating system. However, the potential flow temperatures are low and therefore the use of radiators would not be efficient requiring over-sized radiators. Low flow temperatures are better suited to under-floor heating, where the large area of rising heat provides a comfortable environment that, in conjunction with good insulation below any ground floors and the high thermal mass of the floor, can give good internal temperature control even with large changes in external temperatures.

Another carbon-efficient approach is to use combined heat and power units that obtain the heat for the space heating and hot water from units that generate electricity. This technology can be used to gain efficiencies in burning oil and gas or with biofuels. The use of biofuels is being based on power units such as the Stirling Engine, which, while the principle has been established for many years, is proving effective for these innovative applications. These units are available for individual dwellings, but larger units for commercial buildings or small communities help to make the most use of both the heat and the power that is produced simultaneously.

New fuel sources that reduce the emission of CO_2, limit the reliance on oil and gas, and help to maintain security of supply. Biofuels are being developed from animal waste (biodigesters) that produce methane to be burnt as a gas, normally to produce electricity to be supplied to the national grid although they may be used for small community heating from combined heat and power units in rural areas where a supply of animal waste is close by. This has the additional benefit in that, while it releases CO_2, it stops the release of the methane that is a more potent greenhouse gas than carbon dioxide. Fuel cells that are powered by hydrogen are also being developed for heating units. It is not clear what fuels may be available for the energy requirements of buildings in the future. However, the precautionary principle to seek convergent solutions that both seek to reduce energy demands and look to reduce CO_2 and other pollutants from renewable secure sources should be significant criteria in all new development and the upgrading of existing buildings.

7.5 Water and waste

The building as resource consumer and polluter goes beyond energy use and global warming. Another natural cycle that building development affects is the water cycle. Vital for health, the availability of a good source of drinking water has always been a significant factor in deciding where developments have taken place. With modern supply and distribution systems this direct link between source and development has been broken. Water has provided support for a growing population but, while renewable, cannot have a guaranteed continuous supply if the natural cycle is not understood and respected. It could become another factor that would make the future unsustainable for the next generation.

There are two strains on the supply and distribution system. The first is the use of drinking-quality water for all water use during the life of the building. The need to supply this high-quality water in sufficient quantity for all the uses is becoming unsustainable. The second strain is our disposal of surface water. Our approach has been to take the rain that falls on developed areas and pipe it away to water courses for disposal. This puts the rain back into the water cycle but significantly increases the natural rates of entry of rain into rivers and bypasses the natural evaporation processes and infiltration rates to aquifers, all of which are the main sources of our quality water. These piped services also increase concentrations of water at certain points in the water courses following heavy rain, thereby increasing the risk of flooding. One solution is to slow the amount of rainwater leaving the site of the building through underground storage and water attenuation, e.g. via local ponds.

As well as surface water, development creates foul water that must be piped away for treatment if significant local pollution is to be avoided. Expanding development demands greater infrastructure and again leads to water bypassing the natural water cycle, with treated water being returned directly into water courses. In addition there is a growing acceptance that some surface water, particularly from roads, is polluted and, while not requiring full treatment, need to be disposed of with care if habitats are not to be polluted or heavy-metal pollutants allowed to enter the water cycle.

These issues of the quantity of drinking water supply and maintaining natural water sources along with concerns for flooding have produced several options, most of which address a number of these concerns as they help restore the natural rhythms of the water cycle.

The simplest systems reduce water use in the building with appliances such as dual-flush cisterns for WCs and spray taps for hand washing. The next step is to recognise the variety of water qualities for different uses in the building. While water of the highest quality will be required for drinking and food preparation, water for washing and cleaning need not be of such high quality and water for flushing toilets and external use has even lower quality demands. All water has some minimum requirements for quality, often in colour and particles to be acceptable for use in buildings and must have biological (pathogen) and chemical limits. However, within these limits, recycling and alternative sources of supply are possible.

Grey water systems recycle water that has been used for washing and cleaning to flush the toilets. Some filtering and disinfecting will be necessary, but this system reduces both the need for supply and the volume of foul water requiring disposal. Rainwater harvesting provides an alternative supply of water for flushing toilets and external use (watering gardens and washing cars) by collecting rainwater for use in the building. This reduces not only the need for supply but also the surface water flow rates during periods of heavy rainfall. Both these systems require water to be stored to help ensure supply, but there are likely to be times when demand outstrips this supply, so dual systems are required particularly to flush toilets when the supply of grey or rainwater is not available.

The volume of storage is likely to be large, particularly for rainwater harvesting, where timing between supply and demand is less well matched. Storage provided by tanks will be on or, more likely, in the ground to save heavy loadings on the structure. This will require some pumping, perhaps to a smaller storage tank at roof level, to supply water to appliances on all floors of the building. In addition, when supply exceeds demand, overflow from the tanks must be provided. For grey water this will be into the foul sewer system but for rainwater harvesting this could be to a soakaway.

The reason for the use of the soakaway (rather than a piped drain) is part of the thinking about creating surface water disposal systems that better match the natural run-off characteristics of the development site. These are collectively known as sustainable urban drainage systems (SUDS). The idea is to hold water in catchments to allow infiltration and evapotranspiration (where vapour is released into the air by evaporation from open water and from the ground through plant transpiration) between periods of rain. These systems not only reduce the speed of run-off and help control flooding but also replace water into the aquifers, helping to safeguard water supplies. SUDS employ many techniques to mimic the site's natural drainage pattern, including green roofs that delay water run-off, permeable hard surfaces that allow surface water to drain through car parks and roads, and landscaping and ponds to hold water and improve amenity. Each site will be able to help but SUDS are another example where community schemes over larger areas can be more effective.

Like all community schemes, they need management and maintenance. In the absence of public authority adoption of these schemes, other management, possibly community or commercial, must be put in place for the scheme to continue to be effective throughout its life cycle.

Solid waste

In the same way as water was once seen as being just one quality to be used throughout the building, solid wastes once used to be seen as all the same to be taken away for disposal, normally in landfill or by incineration. Considering solid waste from a building as all the same leads to two difficulties in achieving sustainability objectives. These are the two common concerns of the potential for pollution and resource depletion.

There is limited land available for disposal sites and the waste will decompose, with potential leaching of pollutants into the ground water and watercourses plus the production of methane, the greenhouse gas, into the air. It is possible to select landfill sites to limit harmful leaching, to collect the methane to be burnt for energy production and return the land to support wildlife or development once decomposition is complete, but these options are limited in so far as they are expensive, take time and may not be in convenient locations. Incineration creates potential air pollution and still requires disposal of ash and, again, will face the difficulties of suitable location. Technical solutions are available but, as with responsible landfill, they come at a cost.

Many items identified as no longer required by the building user or operators are made of materials that have the potential to be recycled. However, the processing of materials from waste is very specific to each material, and this leads to the need to see waste as separated waste-streams from the building. Identifying and separating these different waste-streams is best achieved at the point where there is the decision to throw it away. This identification and separation are only of value if collection, distribution and reprocessing are available for the potential of recycling to be realised. In the UK, these facilities are well developed and therefore designing buildings to encourage users to separate waste for collection becomes part of sustainable development. Organic waste can be used for compost, while paper, glass, metals and plastic can be reprocessed, saving on energy, and avoiding depletion of the original feedstock still further. The list of recyclable material is growing but it is still better to reduce and reuse before recycling, but this is a matter for us all in our purchasing and lifestyle choices. However, there is waste associated with the construction of the building, and this is a matter for design and the choice of technology in the specification and detailing of materials for the construction.

7.6 Materials choice and detailing

As the design of buildings reduces energy and CO_2 emissions and promotes the management of water and waste during the operation of the building, the significance of the construction of the building as polluter and consumer of resources becomes greater. This is directly related to the choice of materials but is also influenced by the choice of detailing

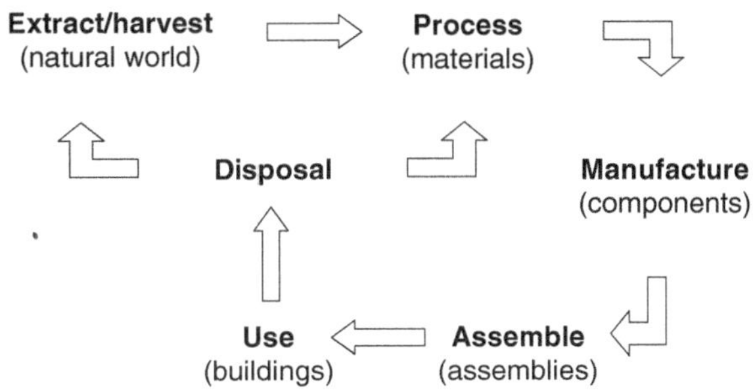

Figure 7.7 Materials life cycle.

and the production options involved in taking the processed material and making it into components to be assembled into buildings.

Evaluating the environmental credentials of materials must consider the life cycle of the material. The stages in the life cycle are shown in Figure 7.7. While we refer to the material's life cycle, it is only in the stages of processing that it can be thought of purely as a material in terms of its environmental impact. When the material is passed to the manufacturing stage, it is formed into components and now a variety of manufacturing processes may be involved, each having its own environmental impact. Simply comparing *in situ* with precast concrete, with their different specification requirements, moulding and transport arrangements, shows that being concrete does not mean that both have the same environmental consequences.

This has led to the development of lifecycle analysis. This sees any part of the building as being traceable through a series of processes back to its source and on to its disposal (or reuse), where each stage has a different environmental impact that is cumulative through the life of the material as it is used in that building. The gates referred to are the gates to the sites where work is carried out on the material to bring it to its component form and then to be assembled into the building to be used and then disposed of. Between each gate there will be transport requirements that have an environmental impact. Local materials with less processing will now be shown to have a lower environmental impact. It does, however, mean that each component in the building must carry an individual analysis. At the time of choosing the specification and details, the impact of the component in use and its disposal will be in the future. For this reason comparative figures are often from source to site gate to consider the manufacturing differences as well as the extraction and processing impacts.

This may still be misleading. Materials are now being specified as part of energy-saving strategies and therefore, it could be argued, have environmental benefits in use. This is the case with concrete being used as thermal mass to reduce carbon release during the operation of the building. It also does not consider recycling potential to conserve resources and reduce pollution. However many stages are considered, the concerns at each stage remain the same. For each process (inside the gates) and for each method of transport (between the gates) there needs to be an analysis of the process resources to identify the effects of pollution and resource depletion.

The focus on carbon as a pollutant has led to the analysis of embodied energy, or more specifically embodied carbon, in materials forming components, transport and even in the assembly process. As with the operation of buildings, issues of water use and waste in each process can be addressed to reduce, reuse and recycle as part of the analysis of the process and manufacturing options. Extraction and materials processing often involve large quantities not only of energy but also of water and can produce some highly toxic by-products that are not evident once they are formed into components and incorporated into the building. These issues can be addressed in the choice of materials and detailing but must be followed through in the responsible sourcing and procurement of the materials and components. Managed sources and local materials contribute towards a lower environmental impact of each component in the building under consideration.

Physical Behaviour Creating Environments: Chapter 8 AT A GLANCE

What? The building is a system of active services and passive fabric whose role and contribution (function and performance) is decided in the design. The aspects of the environment that need to be considered are dryness, warmth, light, acoustics, cleanliness, safety and privacy. Combined, these influence the building occupant's perception of comfort, which may influence their overall health and wellbeing in a positive or negative manner.

Why? Each aspect of the environment that could lead to failure must be analysed to ensure that the performance of the building as an entity, or in large buildings specific zones within the building, perform as intended. This can only be done by understanding the inter-relationships between the various aspects and predicting how the building will be used during occupation. Patterns of occupation will influence the design of the internal environmental quality.

When? The aspects outlined in this chapter need to be considered at the design stage, alongside the need to improve the carbon footprint and lifecycle performance of buildings. The initial analysis and choice will be further refined as the building design is detailed to allow it to be assembled to ensure an interior environment that performs as intended through the year and through various patterns of use. Proposed changes at the construction stage need to be considered for their influence on the overall internal environmental quality before they can be implemented. What may appear to be simple changes can have a significant impact on the performance of internal spaces.

How? For each of the aspects to be considered it is necessary to have a dynamic description of both the external conditions and the response of the construction to those conditions. This can be done by following guidance contained in building codes and regulations, and through predictive modelling and simulations. Post-occupancy evaluation and thermal comfort surveys may also help to inform analysis and choice. Visualising the building as a system is essential to guide analysis and choice.

8 Physical Behaviour Creating Environments

This chapter considers the aspects of environment that are important to the wellbeing of building occupants and their possessions. These are: dryness, warmth, light, acoustics, cleanliness, security and privacy, which combine to create a healthy indoor environment. In achieving these environments the building must be seen as a system employing both active services and the passive building fabric. The overall performance requirements will depend on the severity of the external conditions and the specification for the internal working conditions. The design will have to assign the role and contribution (function and performance) of each part of the construction to the creation and maintenance of the required environment. The range of physical and social environments is shown in Figure 8.1. Once the contribution to performance for any part of the construction is known, it is possible to undertake an analysis of the response of the building under the working conditions to evaluate the chance of failure. For each of the aspects the dynamic response of the proposed construction can be visualised based on an understanding of the law of nature and hence the actions and mechanisms that may lead to failure.

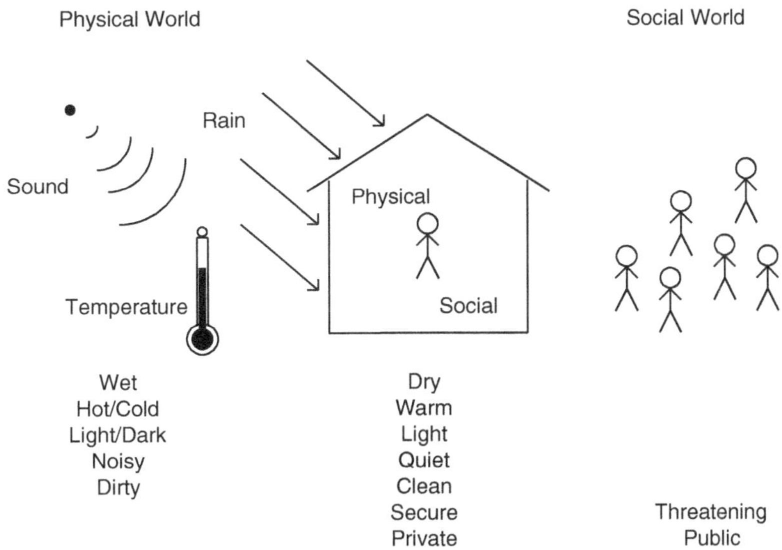

Figure 8.1 Physical and social environments.

8.1 The building as a system

Before evaluating the behaviour of the construction it is necessary to assign functions to each part and define the required level of performance that each part must achieve to ensure the balance of flow and transfers to create the conditions required. A definition of the required environment does not say anything about the contribution to be made by the various parts of the building. Even when the social and physical worlds outside the building have been defined, all that has been established is the total task of modification. Before functions can be assigned to the various parts of the building it is necessary to develop an overall view of how the required environment is to be created, used and maintained. There are three basic groups of technologies available to fulfil these functions:

- Passive fabric
- Active fabric
- Active services

These are shown in Figure 8.2. It could be argued that supply and disposal services are not strictly environmental services, but all contribute to the provision and active creation of the environment, even though they contribute to other operational services as well. Water, gas and electricity are categorised as utility services. Many of the aspects of environment can be achieved with different combinations of these three types of technology, for example the use of artificial lighting to supplement natural lighting provided by windows or roof lights together with the colour of walls. Decisions on this balance are influenced by initial

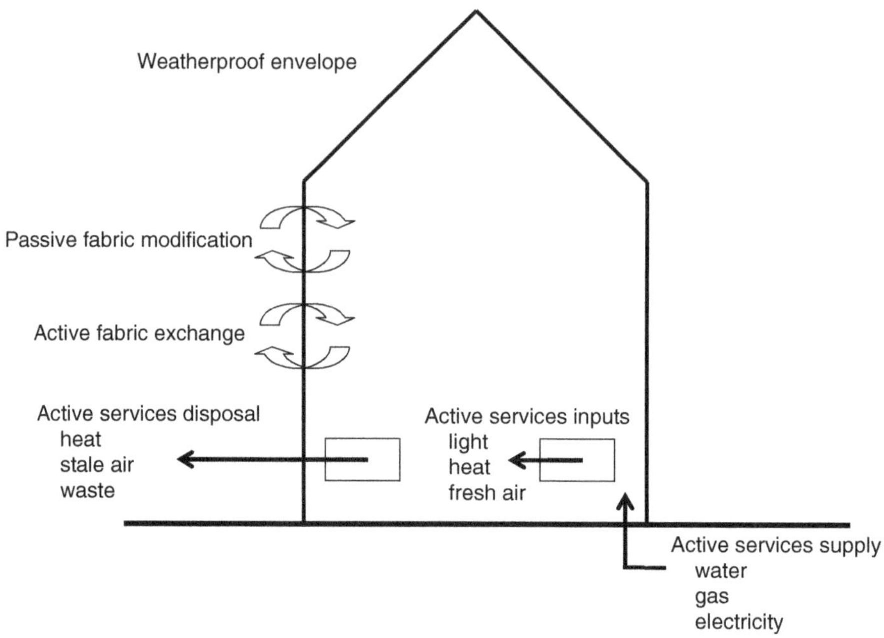

Figure 8.2 Creating internal conditions – possible technologies.

costs, running costs, energy use and the controls to be incorporated. The combination of building design and active fabric solutions can reduce energy use from the active services.

Even when this balance between services and fabric has been established it is still possible to assign the functions of the fabric in several different ways. Some functions are obvious, such as the roof covering's function of weatherproofing to create and maintain a dry internal environment. For others, alternatives exist. In providing a warm environment thermal insulation may be chosen to maintain internal room temperatures to limit the need for heat inputs. This can become a function of either one, or all, the main elements of the building, roof, walls and ground floor, each with various levels of performance contributing to that part of the strategy for maintaining internal temperature levels. When the balance between services and fabric has been established, it is possible to start to develop details and specifications. The approach to each of the three types of technologies is different, generating a different set of decisions for each.

Passive and active fabric

Where the creation of the internal environment depends on the contribution of the fabric, the process of choice is focused directly on the variables identified in Chapter 4. If passive design and fabric is to be employed, then the primary concern will be for:

- The properties of the materials
- The dimensions, usually thickness of materials
- The joints and fixings

For example, if sound transmission is being considered and mass is to be used to achieve performance, the density of the material will determine the thickness required, so long as the joints do not support flanking paths, and the fixings provide sufficient rigidity. Where active fabric such as windows are being used, the focus becomes the joints and fixings to ensure the operation of the moving parts. The material, shape and dimensions of the active component will be determined primarily by their strength and reliability in use, although they may also have a passive contribution to make that will require analysis. They will need to be considered for appearance, as these components are almost certainly visible. For active fabric there will also have to be some form of manual or automated control so that their effect can be brought into use when required.

Active services

The final choice of components and materials in services systems is not as straightforward as the direct choice that associates the specification of the fabric with the environmental conditions. Their choice has become a design discipline. They are systems within the system of the building. Figure 8.3 suggests a way of considering the active environmental systems and their integration into the overall building design.

Outlets for services are required to link the user with supply and disposal systems. Taking a heating system as an example, a radiator may be provided as an outlet but must be linked to a heat source (e.g. a heat pump). In this case the active distribution would be via heated water contained in pipes, most probably using a pump to provide a mechanical force. The radiators would be in each room controlled by thermostatic

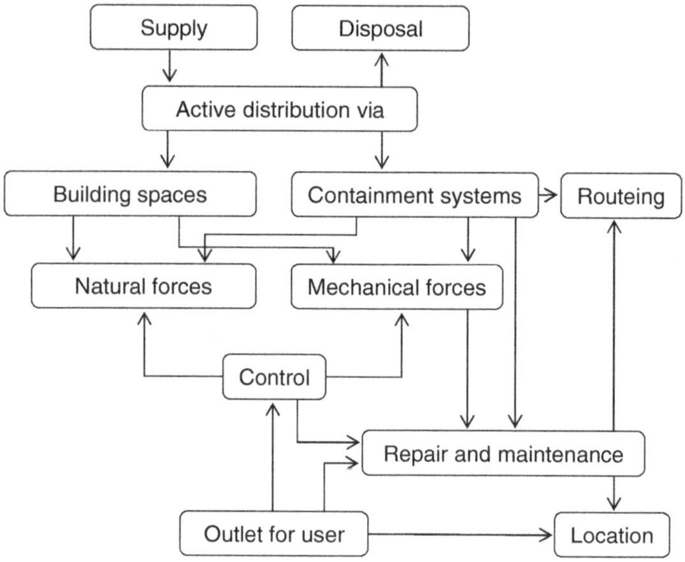

Figure 8.3 Aspects of active services systems.

valves. The heat source to provide the heat would need to be located to make use of connection to an electricity supply, along with consideration of its appearance and as a possible noise source. The pipes may be routed through the floors and either within or on the walls of the building or through specifically designated ducts or voids. This would have to include an access point to allow for repair and maintenance. A possible improvement would be to move away from water filled radiators (wet or hydronic systems) and consider air-based systems connected to heat pumps. These can both heat and cool homes, allowing better control of temperature in cold and hot weather. In another example, a natural ventilation system could use the building space as the distribution system, using natural forces of wind and thermal gradients to move the air from the user, to dispose stale air at high level and bring in fresh air (perhaps tempered with heating) at low level.

Figure 8.3 also shows that, like the active fabric, these active services systems require control. When the internal conditions change, the control mechanism will have to be invoked. This control can be manual, relying on the user to identify unsatisfactory conditions and then being offered some way of re-establishing the required environment – for example, opening a window, if fresh and relatively cool air existed outside the building, to rebalance a stuffy or over-warm room. It would include leaving the occupants to switch on lights when they considered it too dark and switch them off when leaving the room. The alternative is to devise ways of providing automatic control, such as the use of a sensor to respond when design limits are exceeded and send a signal to a part of the system, which can start a chain of events to bring the environment back into design conditions. In the example of the heating system in the previous paragraph, this may be a room thermostat that can switch on the central heating system when the temperature in the room (more correctly at the thermostat) falls below a predetermined limit. In the example of the lighting a simple motion sensor can be used to turn lights on and off depending on if the room is occupied or not.

Assigning environmental modification functions

Table 8.1 shows the aspects of the internal environment that may have to be created and maintained. These are each then identified with the major aspects of the external environment that will influence the achievement of modified conditions. The final two columns indicate typical functions that could be assigned to parts of the construction to achieve the modification required. For most of the aspects of modification, both passive construction and active services could be employed with different functions. For example, to create dry internal conditions it is necessary to understand patterns of precipitation, the conditions of ground water, and the presence of vapour in the air. From this, waterproofing can be assigned to the fabric and water disposal to the building services.

Rarely does one part of the construction act alone in fulfilling environmental functions. Elements of the building, such as walls and roofs, may contribute to many of the functions. Elements are assemblies of sub-elements, such as the external skin of a cavity wall or tiling on the roof, where a primary function is often identifiable. Choice often centres on these sub-elements, for they are normally identified with a major component or material where the focus of the functions is also often clear. For example, an external wall probably plays its part in all the environmental functions. Even major sub-elements such as windows have

Table 8.1 The role of fabric and services

Internal conditions to be established	Significant aspects of the external environment	Possible functions of fabric	Possible function of services
Dry	Precipitation Ground water Vapour in air	Waterproofing Vapour control	Water disposal
Warm	Air temperature Solar gain Wind Humidity	Resisting the passage of heat Thermal mass Ventilation	Heating Cooling
Light	Sun path Overcast sky	Transmission Reflection	Room Task Mood Architectural
Acoustic	Natural sounds Pollution	Transmission Absorption	
Clean	Dirt Pathogens Fungal and plant life Animal life Air pollution	Exclusion Filtering Cleanable	Removal Filtering Disposal
Secure	Social threats Act of God Fire	Resistance Separation (barriers)	Detection Alarm
Private	Cultural and social conventions	Separation (visual and acoustic)	

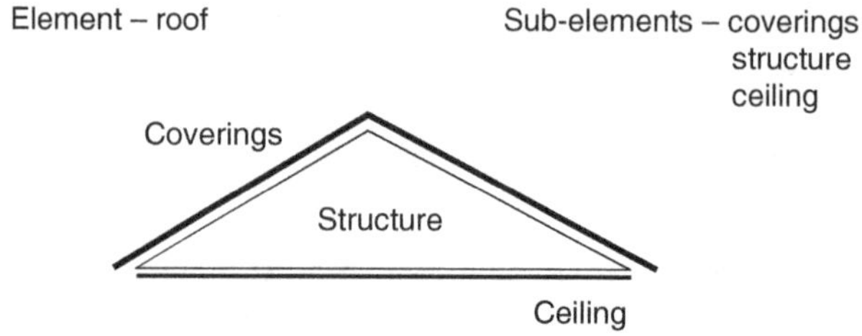

Contribution of sub-elements to environmental performance

Coverings	Dry	Waterproofing
	Clean	Infestation exclusion
	Secure	Fire spread
Structure	None	
Ceiling	Warm	Resistance to passage of heat
	Light	Reflection
	Clean	Dirt exclusion

Figure 8.4 The role of sub-elements of the roof.

a range of functions. Primarily for light and view, these may be used successfully for ventilation but may cause problems with solar gain, privacy and security. As finer division of the sub-elements is considered, down to single components, then function is often clear. Within a wall the damp-proof course (DPC) has a well-defined function of damp control. Analysis at this level must keep sight of the function of the building as one entity, while defining the function of the components and materials of which the sub-element will be made. This division and refinement of the environmental functions is illustrated for a roof in Figure 8.4.

Assigning performance levels for environmental modification

Assigning of a function only identifies the part to be played by an element, sub-element, component or material in creating and maintaining the required environment. Assigning a performance involves defining a level of attainment. While it is unusual for a building's design to be so radical as to cause the functions to be unclear, the new demands of clients and regulations often lead to a change in performance. All the aspects required to set a performance level have already been introduced:

- The range of what will be considered normal external conditions and the limits of extreme or accidental conditions
- The ideal internal conditions and the limits that would still be considered satisfactory, together with control mechanisms that are intended to be included
- The proportion of the task of creation and maintenance of the environment to be assigned to that part of the construction

Thinking through the conditions

The analysis of each aspect of the internal environment will be considered in turn in this chapter. Focus will be on the analysis of the fabric, although the choice of services systems will be introduced. Each section will consider:

- The origins and nature of the environmental conditions
- The nature of the failure linked to performance
- Possible types of solutions
- Analysis of the response involved to achieve performance

A dynamic description of the environmental conditions to be modified and the mechanisms that determine the response of the construction to the expected environments in which it will have to operate should be carried out. This should be based on a visualisation of the flows and transfer taking place through the construction.

8.2 The dry environment

Substances exist in a solid, liquid or gaseous state. Unlike nearly all other substances, water can exist in all these states within the normal climatic range of temperatures and pressures. This creates problems for buildings trying to maintain dry environments. An excess of liquid water is undesirable, but a limited quantity of gaseous water (vapour) in the air is important for comfort, providing levels of humidity. In visualising the behaviour of construction, the possibility of a change of state must be considered. Perhaps the most important example of this is the change from vapour to liquid in the depositing of condensation.

As well as having an undesirable effect on health and comfort, water is a major source of deterioration of construction materials. Detailing has the dual function of protecting the internal environment and the fabric of the building. In this respect, visualising the dynamic interaction between water and the construction is common to the analysis of both environmental control and durability. However, the analyses of the two need to be considered separately since different actions of the water become significant, as, for instance, in the change of state from liquid to solid with frost action, significant for durability but of little concern in considering environmental conditions.

If the requirement is for a dry interior, failure occurs when water enters the building through the external fabric and openings, or from failure of the building services, commonly described as a leak (of water). However, detailing of buildings is also concerned with excessive moisture in the fabric, known collectively as damp. The dividing line between these, both as observable actions and in the appropriate detailing, is often fine, so they are often considered together. Maintaining a dry interior (absence of dampness rather than lack of humidity) requires the control of moisture as both liquid and vapour. As a liquid, the source of water is normally external and hence the external fabric has a major functional role as a barrier to water ingress. Water vapour is predominantly generated within the building; thus the building fabric and ventilation are used to prevent condensation.

When considering the analysis and choice of the fabric to ensure a dry interior, it is usual to see three major threats:

- Penetrating damp
- Rising damp
- Condensation

Among others are wind-blown snow entering ventilated openings and subsequently thawing and leaking.

To visualise the behaviour and analyse the ability of the construction to fulfil the function of damp control requires knowledge of the origins of the moisture, the forces acting on it when it encounters the construction and the pathways then available for it to reach internal surfaces. Predictive methods are available to model vapour and temperature patterns to assess the risk of condensation. Penetrating and rising damp assessment is still largely a matter of detailing to maintain barriers to the progress of dampness into a building by applying tried and tested methods. Testing of prototypes to wind and rain penetration is possible, particularly associated with components systems such as windows and cladding systems. This prototype testing is especially significant when new principles such as rain screening are applied because predicting the behaviour of water can still be a matter of trial and observation. Each of the major threats will be considered in turn, with descriptions of behaviour, identification of pathways and of the major detailing techniques that can be employed.

Penetrating damp

Rain is the major source of water that interacts directly with the external fabric of the building, creating the possibility of penetrating damp via three major actions:

(1) Water that comes into direct contact with the outside of the building will soak into permeable construction where pore networks form continuous capillary pathways through the material.

(2) Water running over the surface of the building is primarily influenced by gravity moving it down the construction, but other forces also act on this surface water, particularly when it meets joints and projections. Water may be drawn into small cracks by capillary action. Wind that accompanies the rain will create pressures around the building, affecting rates of movement of the water across the surface and penetration of the fabric, particularly at open joints. Visualising this complex set of forces acting on water on the wetted surfaces of buildings is particularly important at the detailing of joints between components and at junctions between elements.

(3) Most of the water that falls on the surface of the construction will arrive at an edge. If it is a free edge, the water will fall off mainly under the influence of gravity. However, its behaviour at these edges will also be influenced by wind and surface tension drawing some of the water back under horizontal overhangs. It is important to make sure this water is removed from edges effectively so that it does not have the chance to accumulate and soak the construction more than that expected from rainfall and snow.

The distinction between the first action (soaking in) and the second (surface water interacting at joints) is not so clearly defined. Most construction, such as roof tiling and brickwork, is not monolithic. It is not just one material where the properties of the material will be the only influence on the rate of soaking in. Although they may be considered as one skin, they are made up with joints, which in the case of brickwork are made with a second material: mortar. The behaviour of soaking in is now dependent not only on the permeability of the brick and the mortar but also on the reaction of the water at the joint. Some of the forces identified in the second action (interaction at joints) come into play. In brickwork with cement mortar drying, shrinkage can break the bond between the brick and the mortar to create fine capillary paths. Water penetration can be observed at the joints long before it penetrates the brick.

The depth of penetration of the water soaking into the material depends primarily on two things: the cycle of wetting and drying, and the permeability and thickness of the construction. Prolonged periods of wetting with only short periods of drying in between pose the greatest threat. If the construction is not specified thick enough, or barriers to halt penetration are not provided, moisture will appear on internal surfaces. The three basic construction forms to overcome this direct transfer through the walling material are shown in Figure 8.5.

All these forms can be exploited. At the beginning of the twentieth century most house construction in England would have had solid masonry walls, but by the end of the century the changes in materials, particularly the use of cement mortars, had led to the almost universal adoption of cavity wall construction.

The water that does not soak in, or is not drawn in, will move down the face of the building. If the surface is impermeable then a considerable quantity of water can accumulate. There is a danger of damp penetrating at any time when the water meets joints or projections. The expected pathway and potential quantity of water must be visualised for each situation of joint and projection. The action of capillarity at a filled joint has already been identified between the brick and its mortar if the bond breaks and fine hairline cracks are formed. This is the scale at which the mechanisms must be visualised. Pathways through the fabric are often very fine and tortuous, with damp appearing on the inside, some distance

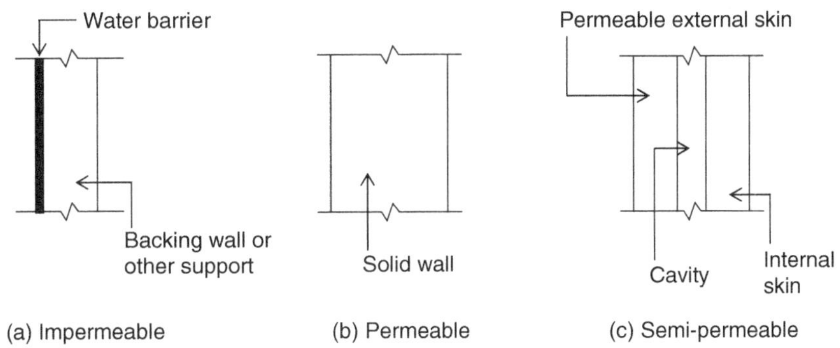

Figure 8.5 Generic forms of walls to resist penetrating damp.

Table 8.2 Forces potentially assisting water to enter at joints

Force	Direction of action	Effect
Capillarity	Any	Water travels through fine cracks and pore networks
Gravity	Downwards	Water enters any opening or crack with downward path
Pressure differences	Any	Above ground wind • Driving rain • Holding water in joint • Moving towards lower pressure • Pumping behind flashings Water pressure below water table in ground
Kinetic	Rebound	Localised to area of impact
Surface tension	Any surface	Movement at underside of projections Holds film of water on surface and on ledges

from its entry on the outside. Other joints are open with no filling material. If these joints are narrow, they may support capillarity if a fine continuous open gap is formed. However, two other forces will assist water penetration at gaps or joints: gravity and wind. These may be significant even where the gap size would not support capillarity. Not only may wind change the surface water's movement directly, but also the pressures that it builds up around the building may blow water through joints or even pump water up behind detailing, such as flashings.

Two other forces that may act on the water when at joints or projections are kinetic and surface tension. Kinetic energy is present in the falling rain. It is the force that causes it to bounce on hitting the surface, so apparently sheltered areas may become wet. Kinetic effects are relatively weak, but surface tension has the potential to cause problems at overhangs and projections. Surface tension may hold water on the underside of the projection, where it will migrate towards an apparently protected joint and enter from below the projection. The basic forces acting at joints are shown in Table 8.2.

Penetration at joints can be aggravated when they occur at projections such as sills. Water will collect on the top surfaces of a projection, causing high levels of soaking or additional water to be drawn in through joints. Identification of the damp pathways and the detailing of damp-proofing at the sill of a window are shown in Figure 8.6. This figure illustrates the use of jointing materials (putty), filled joints (water bar), barriers DPC and, on the underside of the sill, the simple device of the groove or drip detail. In the drip the water collects and gains sufficient mass to break the surface tension when droplets fall from the underside before reaching the joint under the sill.

The distinction between absorption from prolonged wetting and direct penetration through joints and projections has particular importance when thinking about aspects of external climate and rainfall in a particular locality. Any building subject to occasional heavy driving rain may need to be detailed differently to one subjected to prolonged drizzle, even though the total average annual rainfall may be similar.

Finally, it is necessary to consider the water that is shed from the building, particularly where the roof discharges water at the top of walls. If not intercepted by a gutter, gravity will take it towards the ground. It is not a direct threat if it falls directly to the ground, but it may fall or be blown onto the surface of the construction, soaking

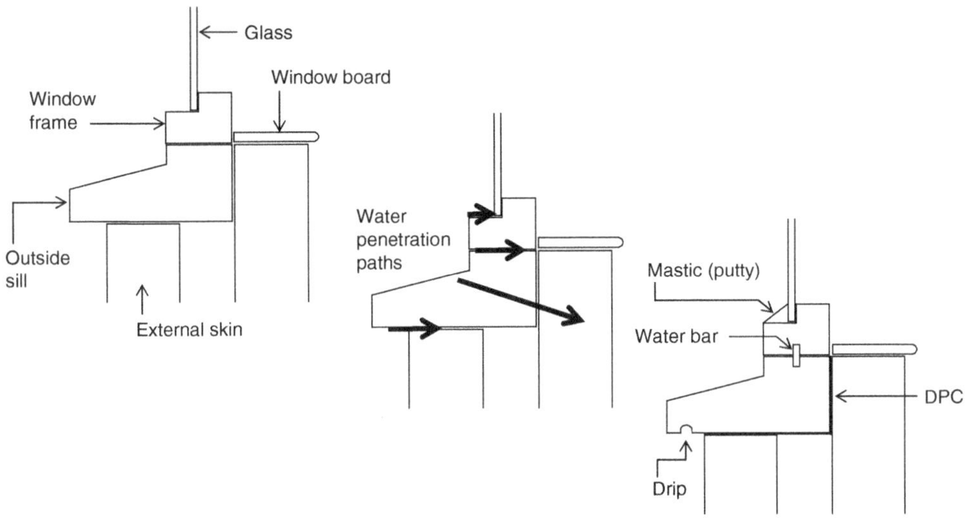

Figure 8.6 Pathways and details to resist damp penetration at sills.

materials to an extent not envisaged by direct exposure to the rain. If it reaches the ground without contact with the construction, it may still splash (kinetic forces) onto the fabric at ground level and increase wetting and soaking of materials close to, but above, the ground. The rest of the water reaching the ground will either directly soak into the ground or collect and flow away on any flat, hard surfaces. Intercepting this water at the edge of the roof is normally the function of gutters, rainwater pipes and surface water drains.

Rising damp

The threat from rising damp is ever present in temperate climates. Moisture is found in the ground most of the year. While the greatest threat may be during wet periods when the moisture content of the soil will be high, the moisture persists in all but the surface layers, even in dry periods. This is even above the water table. If a building is built below the water table, and comparatively few are, then the building is under an additional threat of water pressure assisting entry through the fabric of the building.

The major action associated with rising damp is a transfer of moisture up into the building fabric against the pull of gravity. Water will travel through the fine network of connected pores in permeable materials by capillary action unless a barrier is used to stop it. The usual detailing solution is to provide an impermeable barrier known as a DPC in walls and damp-proof membrane (DPM) in floors. Identification of the damp pathways and the detailing of damp-proofing at the junction of the ground floor and the external wall of a house are shown in Figure 8.7. Given that the major action is capillarity, which exploits the finest pathways, this indicates that any lack of care in detailing or assembly can have a disproportionate effect. Once damp-proofing is breached, the materials on the dry side absorb the moisture and transfer of the moisture will continue to internal surfaces.

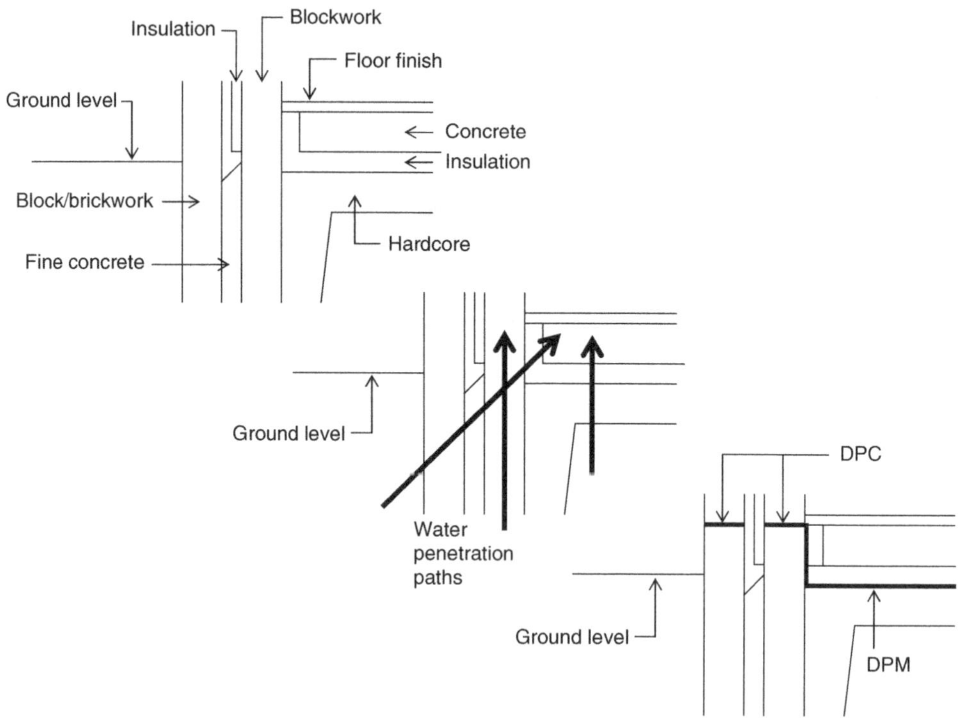

Figure 8.7 Pathways and details to resist rising damp.

Condensation

Moisture for condensation is generated within the building through everyday activities such as cooking, bathing and breathing. This vapour needs to be carried away by any process of ventilation to prevent condensation on surfaces and within the building fabric. Vapour is not a threat to the dryness of the environment (although excessive vapour causing high humidity can be uncomfortable); if all the vapour can be held in the air it causes no damp problems. There is, however, a limit as to how much vapour can be held in the air, dependent on several factors, primarily temperature. If vapour increases at constant temperature or temperature decreases with constant vapour the result may be condensation or the appearance of liquid water. This is most seen on the internal surfaces of window glass, which will probably have one of the lowest surface temperatures in the room. Another vulnerable area is at the internal corners between wall and ceiling and behind furniture where ventilation has least effect. Here the surface temperature may be higher but so will the vapour concentrations. As these surfaces, unlike the window glass, are permeable the condensation will soak in, and the first sign of a problem may be mould growth on the surface.

The temperature at which condensation occurs varies depending on the quantity of vapour in the air. The temperature at which condensation will occur for a given air–vapour mix is called the dew point temperature, where the air is said to be 100% saturated or to have a relative humidity of 100%. If the temperature continues to fall (or more vapour tries to enter the air), more condensation will be produced, as the air can be no more than 100% saturated with vapour.

Condensation can be a problem within the building fabric. Most construction materials are not only water permeable but also gas (and therefore vapour) permeable. The two are, however, not directly related. Gas permeability of a material is a different property from water permeability. When air inside the building is warm, it is probably holding more vapour than cold air outside. This difference is sufficient to cause vapour from inside to migrate through the construction towards the outside. As it is warm on the inside and cold on the outside, heat will also be flowing out across the construction, the temperature of each layer of material becoming colder towards the outside. This drop in temperature becomes particularly large across any insulation. Insulation material does not normally hold up the vapour being transferred, so this may create the conditions for condensation to occur within the construction. Because it happens inside the materials, it is known as interstitial condensation.

Water is now present within the construction. If vapour continues to flow and temperatures remain low, the quantity of condensate increases. The presence of water within the materials may threaten the durability of the materials and may change other properties on which the function is dependent: principally their insulation and air permeability values. Depending on the detailing and where the condensation occurs, it may create dampness on the internal wall. This will depend on there being a pathway for the condensate water back towards the inside face of the construction. Similar forces and mechanisms create these pathways, as discussed in penetrating dampness. If it does not find its way into the building it will need to dry out from the outside surface, where wind, air temperature and the sun will dry the surface, creating the conditions for a migration of absorbed water to the surface and or evaporation. This may be a slow process, and condensate can remain within the construction for a long time. Detailing, including the position of the insulating layer along with ventilation and vapour control layers, can limit the risk of condensation and this is discussed in the section on the warm environment.

8.3 The warm environment

In temperate and cold climates the function of the building is to create and maintain a warm environment. In warm and hot climates the function may be characterised as providing a cool environment, and this will lead to very different strategies and will therefore need a different analysis to the one described in this book. This section will focus on the temperate climate winter heating condition, although the summer condition will also be considered.

Feeling warm is not only related to air temperature. It is dependent on radiant effects, cooling air movement (draughts or breeze) and humidity. This is described as thermal comfort and is influenced by one's age, general health and levels of physical activity.

Radiant effects are a process by which heat is lost (or gained) from the body by radiant exchange. This is independent of air temperature and only requires a cold (or hot) surface to be in direct line with the individual (or another object) for heat to be exchanged.

Areas of glazing can be associated with these effects. The rate of radiant exchange is proportional to the temperature difference, and this is likely to become large if the glazing is directly exposed to sunlight (heat gain) or a clear winter night sky (heat loss). It is likely to be uncomfortable if activities need to be carried out too close to the glazing. Under these circumstances some form of shading, shutters, blinds, or curtains may have to be used, either externally to keep the surface at air temperature or internally to cut out any direct radiant exposure from the surface to the individual. Placing warm radiators beneath cool windows will also help mediate the radiant exchange experienced by the individual if the temperature of the cool glazing is not too extreme. Working very close to a radiator can also be uncomfortable due to radiant exchange. Air movement will come from ventilation, where if the rate of airflow is too high a cooling effect will be felt by evaporation from the skin. This will be discussed in the section on ventilation below. High humidity reduces the efficiency of skin-cooling evaporation, making us feel less comfortable in high air temperatures even with high air flow rates.

The function of the passive fabric of the building in maintaining a warm environment within the building is complementary to the function of the active environmental services systems. It is necessary to see any thermal design to be integrated between the fabric, heating and ventilation arrangements of the building. This strategic decision is also concerned with issues of energy use, clean air and running and maintenance costs. Whatever the overall contribution in each case the fabric will fulfil the same function, only the level of performance will vary.

In a temperate climate the usual concern is the loss of heat through the fabric of the building under winter heating conditions. The more heat that is lost through the fabric by transfer and ventilation, the more heat must be provided by the heating system. This reflects the wider concern associated with energy consumption and running costs. There is also increasing concern with larger commercial buildings regarding cooling in summer, dealing with high internal heat gains and reducing solar gain so that less cooling is required. The role of the fabric under these conditions is different from the winter heat retention role. This will, for instance, require greater rates of ventilation, the use of shading on the external face and exploiting thermal capacity rather than lower ventilation rates and insulation of the building skin required for winter conditions.

To consider the external skin of the building it is necessary to understand how the construction responds when a temperature difference exists on either side of a wall, roof, or floor. The fabric will almost certainly consist of various layers performing different functions, only one of which may specifically be to enhance thermal performance. However, it is the behaviour of the fabric that needs to be considered. It is possible to visualise the materials and layers of construction from the point of view of either ease with which they transmit heat or the resistance they offer to the passage of heat. Generally the description that follows will think in terms of layers offering resistance. The function is resisting the passage of heat. Layers with high thermal resistance have materials normally thought of as insulating materials.

A simple view of fabric heat loss

The simplest view of heat loss through the fabric considers the system of inputs and loss to be in equilibrium. It assumes the temperature on each side to be constant and the rate of transfer to have reached steady state. It also assumes that all heat moves directly through the fabric from one surface to the other. The transfer of heat through the fabric is, under these circumstances, dependent on the resistance to heat transfer offered by the layers of the materials chosen, including any cavities, plus the layers of still air held at the surfaces of the construction. Steady state is the simplest way to visualise the heat flowing through the construction and it illustrates the fundamental influence of the construction. Given a temperature difference on either side of the fabric, heat will flow from the high to the low temperature. The quantity of heat that passes through the fabric in a specified time, the rate of heat flow, will depend on the resistance of the layers, including surface air layers and cavities of air. The greater the resistance of the whole construction, the less heat passes per unit of time. It is not possible under steady state for more heat to be flowing through one layer than another.

The second influence of having layers with different resistances is the way in which the temperature changes across the construction from the warm to the cool side. This is dependent on the relative resistance of each layer. This is illustrated in Figure 8.8. Layers contributing a large proportion of the resistance to heat flow (the insulation) will experience a large proportion of the temperature drop across the construction. Not only is steady state the easiest way to visualise heat flow, but once the resistance of each layer is established it is the simplest condition for which to provide calculations. The rate of heat loss is expressed as a function of the whole construction, combining the effect of all the layers into one figure representing the transmittance of the complete construction. This is a single transmittance value known as the U value.

Heat loss calculated in such a way is based on a description of behaviour that is rarely observed in practice. While the internal temperature may be relatively constant, the external temperature throughout a heating season will vary a great deal. Deciding on a design condition for external temperature becomes crucial to the value of the estimates provided by such calculations. Chosen too low and heating systems will be over-designed for most of the heating season, while chosen too high and conditions of comfort may not be achieved for a large proportion of the days when heating is required. Current design practice in the UK is to take 0 °C as the external design temperature.

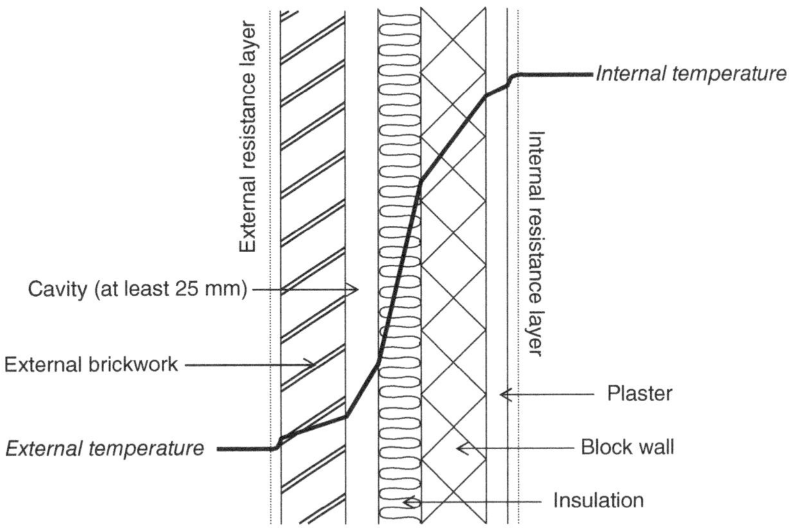

Figure 8.8 Typical temperature drop across insulated cavity wall.

The accuracy of any estimates calculated in this way is sensitive not only to the choice of design conditions but also to detailing. The assumption was made that all heat moves directly through the fabric from one surface to the other. If the resisting layers remain continuous over a large area, this assumption is reasonable. However, whenever the layers are interrupted by structure or openings this assumption must be questioned.

Thermal (cold) bridging

Figure 8.9 illustrates the effect on heat flow through construction where the studs in a timber framed wall panel do not provide a single layer. The studs are said to be providing a thermal (cold) bridge by the diversion of heat, so that it does not pass directly from one side to the other. This will also occur at junctions of floors and around window openings. Heat will flow naturally towards the lower temperature, distorting the temperature profile. This thermal bridging has the effect of losing more than a simple heat loss calculation would suggest. Hence where a building has little continuous high-resisting fabric with frequent penetrations or openings, the detailing to reduce this effect needs to be considered. The potential bridge at the head of a window and the detailing to minimise the effect are shown in Figure 8.10.

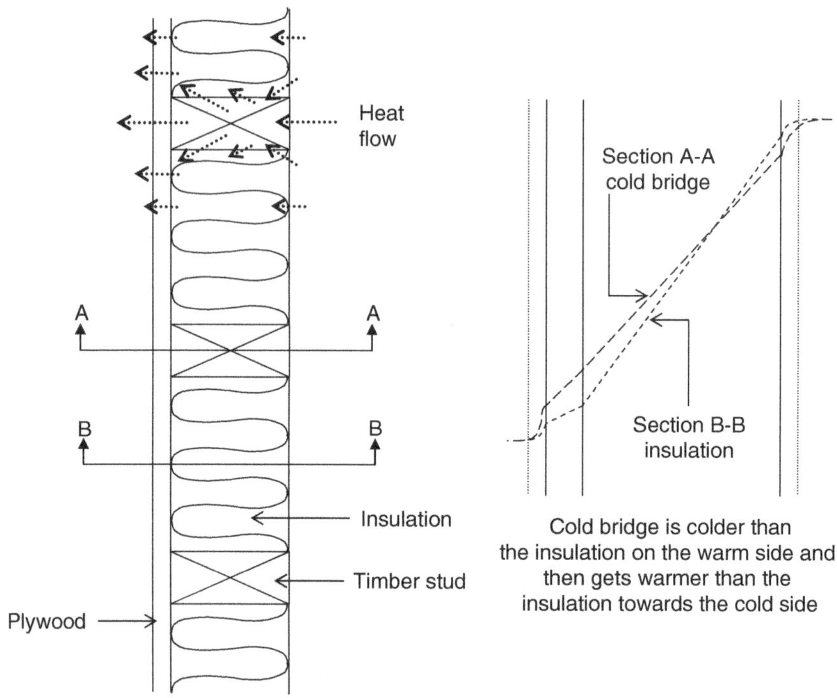

Figure 8.9 Effect of studs as a thermal bridge in timber frame wall panel.

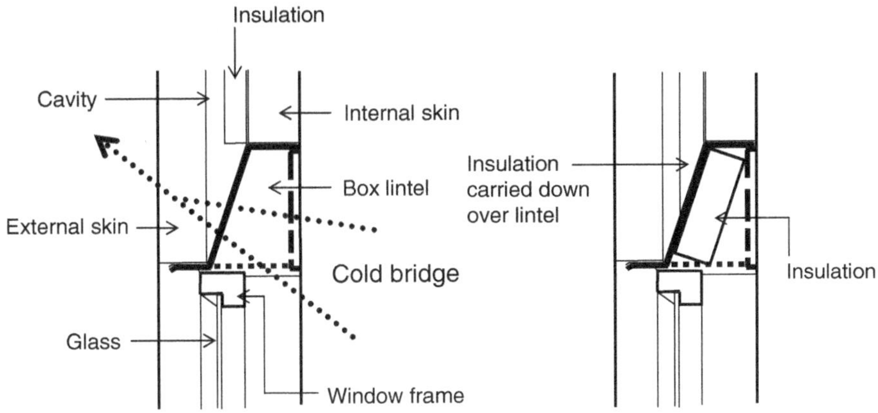

Figure 8.10 Thermal (cold) bridge and details to reduce heat loss at window head.

Non-steady-state heat exchange, capacity and diffusivity

While a building is being heated, the internal temperature will be relatively constant. The outside temperature will vary much more, even on a daily cycle. Not only does this involve the air temperature but also surface temperatures will be directly affected by radiant exchange by direct exposure to the sun (gain) and to clear night-time skies (loss). Surfaces, particularly if dark in colour, will absorb heat if directly exposed to the sun and lose heat if directly exposed to a clear winter night sky. Surface temperatures can by this mechanism reach values significantly different from air temperatures.

Buildings may not be continuously heated. Intermittent heating may be necessary to match the occupation pattern. Under these circumstances both internal and external temperatures are changing, often in the opposite direction, with internal temperatures rising just as external temperatures are falling. To visualise this dynamic thermal behaviour it is necessary to think about how rapidly the temperature profile in the fabric changes in response to temperature changes at the surface. The thermal properties of the materials and thickness of each layer will determine the speed with which the fabric will respond. This needs not only a measure of resistance or insulation value of the layers but also one of thermal capacity. Capacity is a measure of the amount of heat necessary to create a given rise in the temperature of the material (or for the heat to be given up during an equivalent fall in temperature). The rate at which the temperature within the layers changes will be quick when there is little resistance to heat flow and little thermal capacity. With increasing resistance or capacity of the layers, changes in temperature within the materials will be slower. This measure of the speed at which the internal temperature profile changes is a function of both resistance and capacity and is known as thermal diffusivity. Generally, materials offering high resistance are low density with large volumes of air entrapped in the materials, while high thermal capacity materials have higher density and hence a lower resistance to heat flow.

Analysis of response and capacity becomes more important than insulation in climates where the large changes occur between night and day, such as in desert areas, rather than between summer and winter, as in temperate climates. Hence, heavier, thermally massive materials are used. This is like the summer cooling designs in temperate climates where the fabric is used to store the heat generated during the day, which is then purged at night. Analysis of response is also important where buildings must be heated for short periods, when thermally light building may be appropriate. Another strategy would be to have a heating system that was chosen with radiant heaters that heat the individual directly without heating the air so that the fabric has little role to play in feeling warm other than reducing draughts.

Position of the insulation

When only considering heat loss and taking the simple view of steady state conditions, the position of the high-resisting insulation layer is of no consequence. When the behaviour of the fabric is being analysed for any purpose where temperature profile or response time is important, the position of the relative resistance offered by each layer may become significant. If the insulating layers are near to the inside of the building (warm side) then the outside layers will have a smaller temperature rise to achieve steady state. In these circumstances not only is steady state approached relatively quickly but also there is little heat stored in the fabric even if constructed with relatively high thermal capacity materials. Often more significant is when the insulating layer is on the warm side, in which case there is an increased risk of condensation with the lower temperatures that arise on the cold side of the construction.

Figure 8.11 shows the temperature profile for walls with the insulating layer in different positions. The converse is, therefore, that insulation on the cold side reduces condensation risk but may increase warm-up periods where intermittent heating is required. If the condensation risk is high, then additional detailing will be required. This will normally take the form of a vapour control layer or the ventilation of cavities. Ventilation can be particularly effective in roof construction.

Ventilation

Maintaining a warm environment is affected not only by the choice and detailing of the fabric but also by the amount of ventilation. The direct loss of warm air from a room under winter conditions accounts for a significant proportion of heat loss. Ventilation involves

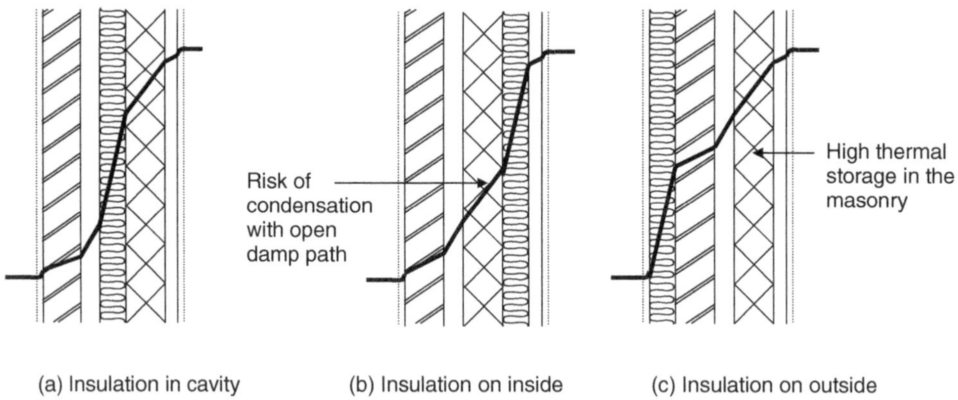

Figure 8.11 Temperature gradient with position of insulation in cavity wall.

air exchange and therefore airflow. If the speed of this airflow becomes too great, it can be called a draught, increasing heat loss from the body so we no longer accept the environment as being warm even if the air temperature has not changed. Under summer conditions the increased speed of airflow may be called a breeze and is welcomed to maintain a comfortable thermal environment. In temperate climates, ventilation is the main means of maintaining a comfortable thermal environment under summer conditions. Air conditioning, being expensive to install and having high running costs, is restricted to a limited range of buildings in these climatic zones.

Ventilation, however, is not just a consideration for maintaining a thermal environment. It also plays its part in two other functions. First, it removes water vapour, which reduces the risk of condensation and hence dampness in the building. Second, it removes stale air, maintaining a clean environment, keeping down the level of airborne pollutants and odours to a safe and comfortable level. This is covered in the section on the clean environment, for there is no doubt that an exchange of air is necessary within buildings, and some means of achieving a controlled level of ventilation must be considered.

For ventilation to take place it is necessary to have two conditions. There must be a pressure difference to move the air, and openings are necessary to let air in and let air out. This assumes a supply of air that is suitable to replace the air in the building. If this is not the case then cleaning by filters or dehumidifying may be necessary, but this will not be considered here. The need for a pressure difference and openings is true for both natural and mechanical systems. The way they are provided, and the means of control are, however, very different. They may be hybrid approaches, the simplest of which is mechanical extraction to increase (or at least control) the pressure difference to dictate the exit opening but leaving the fabric to provide the inlet openings. Even with a full mechanical system, the fabric must be detailed and built so it is not too permeable ('leaky'), as this could influence the performance of the mechanical system.

The part played by the fabric in the provision of ventilation is one of providing openings. These are not only those specified for this purpose such as air bricks or chimneys, but also the cracks that are present around opening windows and doors. Sealing cracks with draught excluders is often seen as a cost-effective way of saving heat. This may be true, but it may also exclude a level of ventilation necessary to maintain a dry and healthy environment. Windows with draught seals should be provided with some other ventilation process such as trickle vents if there is a danger of low ventilation rates in the building.

Pressure differences created by the wind occur naturally around buildings. Typical pressure distributions around a building are shown in Figure 8.12. Air will enter any openings on the side of the building experiencing positive pressure and be drawn out from the side with negative pressure.

A second process that can be harnessed to create natural ventilation is the stack effect. This does not induce a cross flow of air but a vertical air movement. It relies on a temperature difference with warm air at the bottom and cooler air above, but it will be assisted by wind if the upper opening is in a negative pressure and the lower opening in a positive pressure region on the outside of the building. Control of the ventilation is by openings in the fabric, and this control will have to account for a variety of conditions experienced locally by the building. Local conditions can be affected by adjacent buildings, which may have a profound effect on the levels of ventilation achieved in practice.

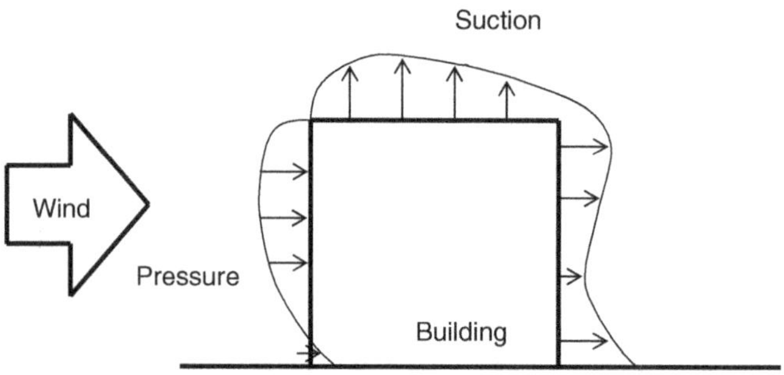

Figure 8.12 Typical wind pressures around buildings.

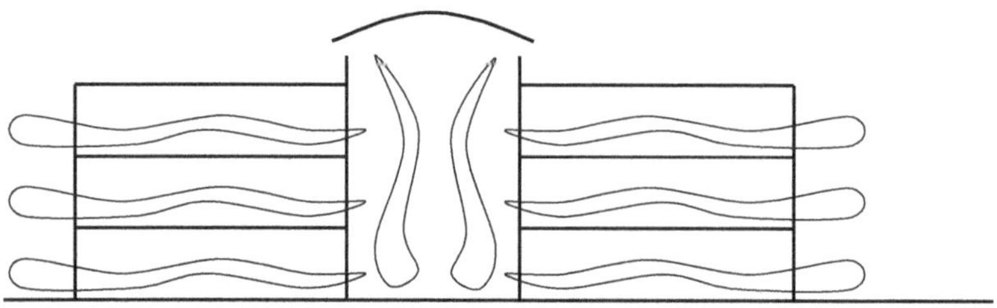

Figure 8.13 Natural ventilation using atrium or internal street.

Visualising these processes demands an analysis of wind pressure and temperature differentials as well as the pattern of opening provided by the fabric. These can be highly complex and are particular to an individual building. Mathematical modelling, CFD simulation, wind tunnel and salt bath trials, can be used to inform and predict natural ventilation to reduce the energy use in a building. The design of the building as one entity must be considered to harness the forces in natural ventilation. One way this can be achieved is with the use of the atrium or street (generally an atrium along the length of the building). The basic design and the natural ventilation forces are shown in Figure 8.13. This design can also take full advantage of natural lighting, again reducing energy use throughout the life of the building.

8.4 The light environment

To fulfil most of its functions the enclosure of a building will almost certainly use materials that do not transmit light. This means that if the user is to make use of natural light, specific consideration will have to be given to getting light into the building. Since most buildings are occupied during the hours of both daylight and darkness, provision must be made for artificial lighting as well as natural lighting. These two may be separate in that the building is to be designed to operate during the day without the use of artificial light. Alternatively there may be an element of permanent artificial supplementary lighting for daytime activity. This has energy implications as well as considerations for the consistency of level and quality of lighting. Over and above providing general levels of light there may be a need for task lighting and illumination to provide a mood or ambience. This may be provided by artificial or natural lighting, depending on circumstances and performance requirements.

Natural lighting

The simplest way to allow light into a building is via empty openings in the external enclosure, but this is unlikely to be satisfactory. Thus it is necessary to use light transmitting materials for enclosure, primarily glass and plastics. Increasing the thickness of these glazing materials reduces the amount of light transmitted, limiting them to relatively thin sheet applications. Glass and plastics have limited strength and are unable to carry structural loads from the building, although they will have to resist wind loads. It is common for the materials to be framed to provide the strength required in the form of a window. As single sheets, the glazing materials will transmit heat easily, giving high heat loss in winter and heat gain in the summer, and will provide poor sound resistance. At the same time they must remain waterproof. The window is often used to control ventilation by providing sections which can be opened. They may have limited ability to provide security and privacy.

These components, such as windows and roof lights, become sub-elements of walls and roofs of which they form a part. They must contribute to the performance of the element as an entity while accepting that they will have different performance levels. These components may also be made of different materials from the fabric that surrounds them, and the detailing of the fixings and jointing to the main enclosure materials must be thought through independently and yet be complementary with the consideration of the elements as an entity. In addition to these considerations the size and shape of windows has a significant effect on the appearance of a building. Windows are integral to the external and internal aesthetics and overall feel of buildings and spaces, adding to the distinctive character of the building.

Against this background the solution to providing natural light may often be a compromise with the other functions of the building. Its importance in function and design terms varies greatly, depending on the use of the building and the designer's interpretation of its importance to the design. The use of multi-layers of glass or plastic components is now common and recent development in materials has given these multi-layer transparent components improved thermal and sound performance, and strength and fire-enhanced properties to help security functions.

Source of natural lighting

Being dependent on the sun the quality, intensity and time when natural light is available are highly variable. These vary with latitude, building orientation, the time of year and weather conditions. In the northern hemisphere light from the south is generally brighter than from the north. It does, however, vary in intensity, often causing uncomfortable or disabling glare rather than even illumination. Light from the north is more consistent and reliable but colder. This made it suitable for factories and warehouses where roof lights on the steep pitch of the characteristic saw-tooth roofline were orientated to face north. For other purposes, variety and qualities of brightness and warmth may be more appropriate, which, with some control for glare, is best provided from the south.

The hours of daylight vary with the seasons, and the light levels and the angles from which the sun can be directly viewed will vary. This will give a quality of light that varies with the seasons but also means that shading may have to be variable to match the angle and direction of the sun. The cloud cover affects the quality of the light. The highest level of light is available with no clouds covering the sun, whereas storm clouds can all but eliminate the light, creating near-night conditions during the day. With all this variety it is unlikely that performance levels can be assured throughout the year. Bright light may need to be shaded, while storm clouds may dictate that artificial light is required. This variety is greater in temperate rather than in hotter climates, where the sun is more predictable and remains higher in the sky. In buildings designed for these hot climates windows may be small and/or be permanently shaded by shutters or verandas facing the main path of the sun to control overheating to the interior.

Variety in natural light is not necessarily undesirable. Changing levels of intensity and colour give information about outside conditions. They keep occupants in touch with natural conditions, marking the passing of the natural cycles of the day and the seasons. In addition the light received brings with it a view from outside. Windows, and to a lesser extent roof lights, allow occupants to look outside to see what is going on. In many cases, landscape (or cityscape) can be designed to provide spaces seen as visual extensions to the internal space. This gives vistas and panoramas that enhance the activity for which the building is designed. The increased cost of offices on upper floors of skyscrapers and penthouse suites is based in some measure on the view they provide.

Despite the technical difficulties of providing the performance of the enclosure where lighting openings are to be included, and the probable additional capital and running costs, windows and roof lights are judged to be good value (if not essential) in the provision of views, light and the influence they have on the appearance of the building. This makes the technical challenges particularly acute.

Penetration of natural light to internal spaces

Given the prevailing external lighting conditions, the quality and quantity of light reaching the internal space are controlled by the size and shape of the openings, the opacity of the glazing material and the shading or other control mechanisms provided. The proportion of light then available penetrating into the space depends upon the area and shape of the room, both its width and depth. Lighting levels reduce with increasing distance from the windows, but penetration can be enhanced by the careful choice of shape and position of the openings, particularly the height of the head of the window.

Lighting levels from natural light are also influenced by reflections from objects or surfaces, particularly towards the back of the room. Surfaces that are shiny or light in colour will reflect a greater amount of light, raising the level of light at the working surface. In the same way as it is possible to provide shade from the direct glare of the sun at the window it is possible to provide reflectors to enhance the levels of light incident at working positions within the room. Predicting lighting levels from natural lighting is now well established. Data on the external conditions and calculations that take window shape, depth of room and the reflection of surfaces into account are available to predict the level and reliability of natural lighting into a building.

Artificial lighting

Artificial lighting will be required during the hours of darkness and to supplement natural lighting in large spaces and spaces that may be designed without windows for security. Windows left uncovered will then act to transmit the internally generated light out and the view will then be available from outside of the inside, affecting privacy and security. Heat loss will also increase as the temperature falls at night. Given these circumstances, curtains, blinds and shutters become associated with window design. Depending on the tasks to be undertaken within the various spaces within a building there may be a need for different levels of lighting, colour and modelling requirements. However these are resolved there is often a need to provide sources of artificial lighting outside as well as inside the building.

In single-storey buildings (or the top floor of a multi-storey building) this can be solved with roof lighting. In very deep multi-storey buildings the use of light wells or atria can bring light to a central space, making light available to areas some distance from the outward-facing walls. In other circumstances artificial lighting will have to be used, although this has energy and running cost implications over the lifetime of the building. For many buildings, the daytime use of artificial lighting is only associated with limited periods when the external conditions are poor. Even if general lighting levels with natural lighting are satisfactory, specific task lighting may still be required and provided by artificial sources.

Sources and control of artificial lighting

The source of natural light is variable. Artificial light is predictable and controllable, and the parameters of quantity and quality can be chosen. Since the case for artificial lighting also must consider energy consumption, it is important to be able to identify the energy use for a given output, which varies between the different sources.

The demands made for lighting both internally and externally vary greatly and hence a wide range of sources has been developed. Originally powered by oil or gas, artificial lighting is now almost exclusively powered by electricity. The initial characteristic of the light is governed by the way the electrical power is converted into light. Whether the light is generated by the incandescence of a solid material such as tungsten halogen lamps or the excitation of a gas such as the fluorescent tubes, it is this conversion process that determines the basic colour and the power consumption of the source. This light can then be harnessed to best effect by the fitting and positioning of mounted units. Directional spotlights get their power and colour from the light source, but their directional capacity comes from the way they are housed and focused, with lenses to minimise scatter, as with theatrical lighting. Colour can also be added to white light with filters.

More general control over individual rooms or sections of large areas is achieved by switching the electricity on or off, or by dimming to reduce the level of light emitted. In many situations this control can be manual by providing switches for the occupants to operate. Many LED lights are now 'smart' with the colour controlled by the user via their mobile phone app. For energy saving, some level of automation should be considered. Often based on movement sensors to identify occupants, automatic switching will turn off lights when rooms are not being used and turn them on again when people return. This may have to be set to leave low levels of illumination so that safety is not compromised. Light level sensors can be used to switch on lights as levels of natural lighting reduce. It is particularly effective to bring in banks of lighting parallel to the windows to supplement the natural lighting at the back of the room without using lights close to the windows where natural lighting is still satisfactory.

As with natural sources, artificial light received at a point in a room is not just in a direct line from the source but also arrives from reflections from surfaces. The strength of these reflections is dependent on how close the surface is to the source and on the texture and colour of the surface. This can be used either to increase the level of illumination (quantity) or to create effect (quality) when used in conjunction with its position.

The predictive methods and the data available for the sources concerning the levels of general illumination are well developed. Both predictive models and data on reflective surfaces and the manufacturers' values for the sources and the luminaires are freely available. Even predictions of loss of efficiency with time, energy usage and heat generation are possible with the current data available. Architectural lighting and effect or mood lighting are less susceptible to calculation and need a measure of imagination, but calculation and computer modelling can give clues to comparative effects especially in terms of power and the interpretation of light and shade for soft or dramatic effects.

8.5 The acoustic environment

Sound can be pleasurable and desired or unwanted and is perceived as noise pollution. Sound that is wanted can be enhanced or degraded by the fabric of the building, particularly by the internal finishes and the shape of the room. Most wanted sound is generated within the space occupied by the listener, most often associated with talking and music; however, sound from the natural environment, such as birdsong, may also be welcome. Like a view or the changing quality of natural light from a window, sounds limit any sense of isolation that can be psychologically undesirable. An increasing number of modern sounds are now considered unwanted, such as the noise generated by traffic on a nearby road, trains and aircraft. They are categorised as noise pollution and as such are now being given much attention in reduction at source, but this still often leaves excessive noise that must be reduced through careful choice of materials and detailing of the construction.

Characteristics of sound

Sound has two characteristics as identified by the ear: frequency (pitch) and loudness. Frequency determines how high or low we perceive the sound to be. If too high or low, it becomes inaudible to humans, although some pets can hear higher frequencies, and some low-frequency sound does affect other organs in the body and causes discomfort, but these sounds are rare. Combinations of frequencies can be harmonious, as in music, or carry information, as with speech. Groups of frequencies that are disharmonies or carry no information are often called noise and, while useful as alarm signals, are generally considered undesirable. Loudness is associated with volume. Because it has energy or power, it has the capacity to cause pain if it excites the mechanism in the ear to excessive movement. At the other end of the scale, quiet noise can be irritating, since it is frustrating if you think it is carrying information that you cannot obtain, or it is continuous and begins to intrude on one's consciousness.

All sound, whether wanted or unwanted, reaches the ear through vibrations in the air. Airborne sound can travel through the construction, and hence the provision of a barrier will only modify and not eliminate all noise. Further, impact sound created by striking the construction, such as footsteps on a floor, will create a sound not only within the room but also transmitted to other areas by the direct connection of the structure and fabric of the construction. The transmission of the sound around the walls and floors directly between source and listener is by means of flanking paths. These noise sources and transmission paths are shown in Figure 8.14. In addition, mechanical plant has the potential to create noise, and this can be passed through the building via the services duct and pipework systems.

Sound created within the space

The sound will travel out from the source and there will be a natural decay of its power as it moves away. If in a direct line with the source, the listener will receive the sound directly. The sound will continue through the air in the room, striking other objects and eventually the floor, walls and ceiling that make up the room. Sound reaching openings in the floor, walls and ceilings will pass straight through. The sound that strikes objects or the surfaces in the room will be partly absorbed and partly reflected, depending on the nature of the surface. The reflected component will return into the room, where it may be received with a reduction in power and fractionally later than receiving sound direct from the source. These reflections will continue to be heard by the listener until the power in the reflection is too low to be audible. The quality of the sound is, therefore, highly dependent on the acoustic (absorption and reflection) characteristic of the objects in the room and the surfaces. However, since the sound loses power with distance, and takes time to return as a reflection, the size and shape of the room are also important. The time each individual sound takes to decay to become inaudible is known as the reverberation time. With highly reflective surfaces in a large room, the reverberation times will be increased.

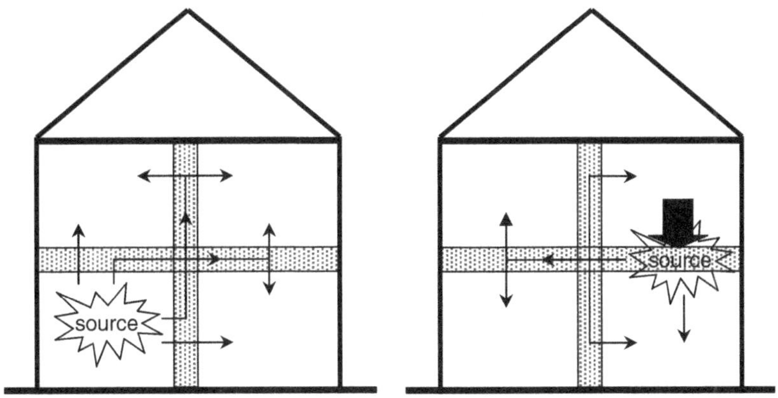

(a) Airborne sound transmission paths (b) Impact sound transmission paths

Figure 8.14 Sound source and transmission paths.

For speech and music it is important to have a quiet background, with little or no noise from outside the room, and little or no generation of sound within the room other than either the speech or the music being listened to. For both it is necessary to have a good strong direct path from source to listener, but for reflections and reverberation time the requirements for speech and music are very different. For speech, a strong early reflection reinforces the sound, but later reflections interfere with the clarity of the separate sounds. Reverberations as low as a twentieth of a second are required. For music it is the later reflections that enhance the sound, enveloping the listener in the sound field. Now reverberations of one to two seconds are required. This enveloping also requires the room to have a certain volume to maintain a smooth and prolonged decay of the sound.

There is one other phenomenon that can occur if a sustained sound with a dominant low frequency is coincident with the dimensions of the room. This is most likely to arise from vibrations from outside the room, which will be considered later, but the characteristic is associated with the room. If the room dimension is a function of the frequency (wavelength) of the sound, a standing wave may be generated sustaining that frequency.

The major influence the construction can have over the quality of the room acoustics is associated with the reflection/absorption of the finishes, having considered the influence, particularly absorption, of the objects (including the listener) in the room. In general terms, hard, flat surfaces are highly reflective. As the surface becomes progressively softer and more textured, absorption is increased. Soft, fissured or even re-entrant surfaces are sound absorbent as the energy in the vibrating air becomes dissipated within the cavities of the irregular surface.

Airborne sound made outside the space

External airborne sound is usually unwanted, but, as has already been pointed out, total sound isolation may not be a sensible performance requirement. The sound will be modified by wind and other atmospheric disturbances. Sound generated in another space will be influenced by the objects it meets, being either absorbed or deflected, and it will impinge on the dividing construction, such as the walls and floors. Openings in the construction will not change the sound; it will enter (or leave) the room unaffected. It is important to appreciate the significance of the size of openings in transmitting sound. Small openings may have a large effect, considerably reducing the sound reduction performance of a barrier, however well detailed.

Walls and floors of the building will act as a barrier to sound. The construction will reduce the loudness but not equally for all frequencies. The action of the construction in reducing the noise is twofold: it dampens the vibrations due mainly to the mass of the construction and it isolates the vibrations if discontinuities (layers of resilient materials or cavities) can be introduced in the detailing. Hence the overall performance of a wall, floor or roof in attenuating sound will depend on the openings, mass and discontinuities and any sound-absorbent materials in the cavities. Sound will also be carried deeper into the building by other elements attached to the initial barrier along flanking paths, making edge detailing particularly important. Figure 8.15 shows possible soundproofing details for a timber joisted floor, illustrating how mass, discontinuities including abortion in cavities and edge details can be integrated into a basic structural floor.

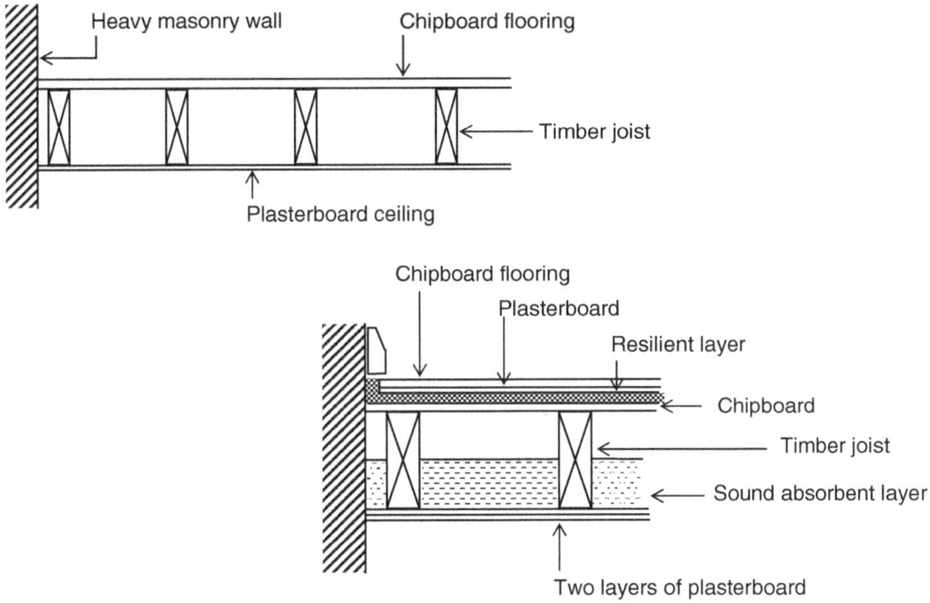

Figure 8.15 Timber joisted floor and details to provide sound proofing.

Care must be exercised in assessing the performance of cavities. If the cavity is too small, it has little effect as the vibration energy is transferred directly across the cavity. This is known as acoustic coupling. It happens in standard double-glazing, where the gap between the glass panes is small and therefore the soundproofing effect is almost entirely due to the additional mass of the second pane of glass. For cavities to be effective they should be around 200 mm wide, or larger.

While the potential sound reduction lies in the basic construction of the floor or wall, the part played by edge conditions and flanking paths means that the actual transmission loss when the building is in use may be less. Transmission loss of the proposed construction can be measured in the laboratory to provide a sound reduction index R_w normally between the frequencies of 30 and 315 Hz. A value of 40 dB reduction will render loud talking inaudible. This value helps to select appropriate specification, but these laboratory-based values may not be achieved when incorporated into the final construction, depending on the detailing and any variability arising from the site operations. It is possible to measure on-site performance, where the transmission loss $R¢_w$ will be quoted.

Sound generated by striking the fabric or from mechanical services

Sound is created by directly setting up vibrations in the construction. These vibrations will be much stronger than those set up by airborne sound reaching the fabric of the building. Flanking paths created by continuity between elements of structure and fabric are important since if the vibrations can travel across junctions between elements and via joints and fixings the noise will penetrate deeper into the building. Mass will again play its part in damping the sound, but rigidity and natural vibration or resonance of the components sympathetic to the frequency of the noise may become important. Discontinuities and edge detailing are particularly effective in limiting the transmissions of these sounds. Striking hard surfaces creates the greatest sound, not only in the room but also in setting up vibrations throughout the structure. Soft surfaces dampen the original impact and absorb the airborne sound that results.

There are problems associated with the pumps and fans of the services. The continuous rotation generates noise, but the movement of the rotation may also set up continuous vibrations in the construction to which it is attached. This may be the fabric of the building supporting the pump or fan or it may be the pipe- and ductwork to which it is attached. Isolation or discontinuities must be used to mount the pumps and fans and connect them to the distribution system.

Methods of prediction

Definitions of a comfortable acoustic environment are well established for a wide range of activity including spaces with both general and specific acoustic demands (e.g. offices and concert halls). There is a great deal of empirical measurement of the characteristics of the materials commonly used in construction, and predictive calculations based on theory can utilise these values to predict performance in a wide variety of circumstances. Modelling and simulation can be used in the design of specific spaces. For the predictions to be fulfilled much care is necessary in the design detailing and the production phase. Identifying isolation discontinuities and in particular edge fixings to elements of the construction are of prime importance. During site operations, fixings and edge details need particular care. Gaps at junctions and the rigidity of connections can be crucial in achieving acoustic performance.

8.6 The clean environment

As with all aspects of environmental performance it is necessary to understand what activities are being carried out in which part of the building. Acceptable standards of cleanliness will have to be defined for the occupants and may need to be considered for the materials and equipment to be housed and used within the building. For all these aspects of the activity, cleanliness will have to be defined against two complementary sets of criteria: those that would be detrimental to health or operational requirements and conditions that would be socially defined as unacceptable.

The socially defined criteria for cleanliness may not be detrimental to health or operational efficiency. They will, however, contribute to the social and psychological health of the occupants. The distinction between the health of occupants and the efficiency of operations is also important. Definitions of cleanliness will have to be identified against ideals of hygiene, the control of pathogens harmful to life or health. They will also have to be defined for the materials stored and equipment operated in the building. Levels of dust and dirt may be detrimental to health and may interfere with the operation of equipment.

Another way of considering cleanliness is to identify the possible undesirable components of the environment as physical, biological, or chemical. The term 'clean' is most associated with the absence of dust or dirt. These particles will be of various sizes and possible composition. They may be inert minerals but are also likely to include matter of biological origins (dead skin cells). They may also be chemically reactive if associated with, say, industrial waste. Particles are most associated with surfaces but can be suspended in the air. The air, however, also carries other potentially undesirable components. Gases can mix with the air and odours can be carried in the air. The source of both gases and odours can be either natural or manufactured, and while some can be a health hazard, some are only a nuisance. Some gases, such as carbon monoxide produced from faulty heaters or the burning of fossil fuels have an immediate deadly effect, while the naturally occurring gas of radon from some ground conditions only shows its effect after many years. Other gases such as methane and carbon dioxide from the degradation processes in the ground such as landfill sites that are now being used for development are not just a health hazard but also introduce the safety risk of explosion. In addition to particles, gases and odours, it is now possible to detect appreciable amounts of electromagnetic flux, the long-term health effects of which are still under investigation.

Surfaces and the air are two major areas of concern in providing a clean environment. Water is also a medium that must be clean. Once it has entered the building at an appropriate quality, avoiding cross-contamination becomes a priority in the design of the water distribution systems. Surfaces, air and water are all open to inspection. Occupants and equipment encounter them. However, within the construction or in unused or not often visited spaces there is the possibility of infestation. Varying in size from cracks in the construction to whole loft spaces, there exists the potential for infestations of animal and plant life. Infestations of insects, rodents or birds are all possible, as are bacteria in some locations. Plant life is most likely to be mould or other fungal growth, as these often do not need light to maintain growth. Infestations of this type not only are undesirable against the notion of a clean building but also can be harmful to the fabric of the building. Control of these infestations is important for the durability of the construction as well as maintaining a clean environment for the occupants and the operations of the building. It is, therefore, necessary to undertake an analysis of the air in the building, the surfaces of the construction visible and in contact with the operations in the building and the infestation potential of the construction and unused spaces in the building.

Clean air

It has long been recognised that particles in the air, particularly from smoke, result in a high incidence of respiratory disease as well as making surfaces, clothes and skin dirty. Similarly it is recognised that dampness in the air is not desirable and that limits to humidity are sensible. Although not as much of a problem as foreseen by early medical predictions, some disease is carried in the air, although often associated with water droplets suspended in the air and breathed in. Smell is another component of the air that, if allowed to concentrate, can become unpleasant, although it is not really a health hazard. It is now being recognised that very low levels of chemical and ion (radiation) concentrations in the air can also be injurious to long-term health or have a low-level debilitating influence. The full implications of many of these are not yet understood and hence safe levels have not yet been established. It appears, however, that as we 'clean' our environment we improve health but then become susceptible to thresholds of other components in the environment that then redefine our levels of performance.

This general process of demanding new levels of cleanliness has also been true where the activity in the building has been manufacturing. This can be seen at its most obvious in the electronics industry, where some processes are now housed in areas called clean rooms.

Clean air is maintained primarily through ventilation. Most harmful components carried by the air are generated from within the building and the process is one of replacing the contaminated air with 'fresh' air from outside. Increasingly the air outside also contains contaminants and, therefore, some measure of processing of either the stale internal air or the incoming outside air must be undertaken. Typically this is the process of filtration with a variety of processes used for physical, chemical and even biological cleansing.

The process of ventilation has been described earlier in this chapter. The necessity of a pressure difference and openings to create the air movement is the same, but the rates of ventilation may be significantly different. It is necessary to identify not only the types of contaminants but also the rates at which they build up and the threshold levels at which the air is to be considered no longer clean. Like all aspects of the clean environment, the performance level of clean is not absolute. There are only acceptable levels or thresholds of contaminants, many of which cannot be detected by sight or smell and require monitoring in other ways. Traditional methods of leaving the occupants to open windows when the room becomes stuffy will still be satisfactory in many situations. However, contaminants such as radon are tasteless and odourless, and therefore must be detected by other means. As a gas from the ground, its control will require gas-proof layers, like the damp-proofing discussed earlier.

Cleanable surfaces

The specification of surfaces is that they shall be cleanable. This demands that the nature of contaminants must be known, as must the means that will be used to clean them and the frequency of cleaning. Considering the lifetime costs it is more than likely that more will be spent on cleaning a surface than on its initial purchase. In the domestic situation, reducing the time needed for cleaning may be seen in relation to its impact on quality of life; to a commercial undertaking this will inevitably be related to time and costs. The methods, machines and substances used to facilitate cleaning will depend on the surfaces to be cleaned and the environmental impact of the substances used.

Infestation

Most infestations are unpleasant rather than a threat to health. Infestations occur where either plant or animal life breeds and lives within the fabric or within voids created by the detailing, or within the furniture and stored items, particularly food. The scale of infestation and the size of the animals vary considerably, as does any definition of what might be unacceptable. Generally, in the developed world, very low levels of infestation are tolerated. Performance specifications would wish to exclude the larger animals such as birds, mice and rats and most forms of insects. Some life sharing a building is, however, inevitable. What is shelter to us provides shelter to many other life forms unless the choice of material and detailing precludes them.

The first line of defence is exclusion of the mature adult. Although this is possible with animals and larger insects, it is often not possible with the smaller species and bacteria. In the case of plant life it is the seeds or spores that must be excluded, and this can prove very difficult. Openings that lead to internal, dry cavities must be detailed and constructed with care to ensure minimum dimensions to exclude adults. Larger openings must be protected with screens to prevent infestation. If exclusion is not an option, the key to stopping infestation is to understand the life cycle of the potential infester and the conditions necessary to sustain each stage of its life. Removing food sources may deter an infestation, but for most infesters it is shelter and protection that they gain from the use of the building, which allows them to breed and hence establish an infestation. If it is difficult to limit the conditions suitable for habitation. While this may be part of the specification, as with wood preservatives, it is more likely to be arranged by the user during occupation of the building.

So far, infestations and breeding patterns have been considered. In some cases it is necessary to exclude adults from rooms, not because they would cause an infestation but because they would be a nuisance or health hazard. This would include flying insects. Similar detailing is necessary, although this may be contrary to requirements for ventilation because screens put over openings will compromise ventilation.

Our knowledge of the animals and plants that are likely to seek shelter within our building is well developed and much traditional detailing takes the elimination of infestation into account. However, we must be ever vigilant in that poor detailing, poor construction and inadequate maintenance may lead to colonisation and infestation. This may be an unintended consequence of using new materials and methods.

8.7 The safe environment

The need for safety in a building arises from the perception of a threat and security. Understanding the nature of the threat determines the performance required of the construction, the nature of failure and hence the choice of materials and detailing. Threats can be seen as arising from natural events, fire and from within society.

Natural threats

Many of the natural threats have been discussed in other sections of the text. Natural threats not yet discussed come from what is often termed an act of God. It includes the natural extremes of climate, such as high winds and excessive snowfall, and other natural occurrences, such as earthquakes. This threat can be identified with the performance of stability and strength. These natural threats exert forces that the structure and fabric must withstand without failure. Structural analysis and design are now well established, even though the natural forces involved are often complex and dynamic. The dynamic nature of the forces means the solutions may have to be carefully worked out in terms of the response of the structure and fabric of the building. As the planet warms the air can carry greater amounts of water vapour, which is resulting in heavier rainfall, often in shorter periods of time. This may result in flooding from surface water runoff and overflowing rivers. Practical steps can be taken to hold back water inundation if the choice of construction fabric is limited. Infrastructure projects in coastal and river defences for low-lying areas may limit the likelihood of flooding but it may not entirely remove the threat. Location of the site and the initial choice of finished ground floor levels above predicted flood levels for the development are the most effective measures to take.

The threat of fire

The source of most fires in buildings comes from the activity within the building or a malfunction of the (electrical) services. It is also associated with possible threats from society when related to deliberate damage via arson. Fire is a major safety issue in the built environment and its analysis concerns the whole building. It has an influence on most aspects of the building design, construction and use. There is much regulation concerning fire safety design and this reflects the level of threat that is perceived in buildings. Fire is a major threat to life and has a financial implication with the destruction of property. Therefore the design of a building must take both life safety and economics into account.

Many of the requirements for safe construction are now laid down in codes, standards and regulations. Specialist input on fire engineering and fire safety will be required for all but the simplest and smallest of building designs. Fire engineering techniques use computational simulation and modelling of smoke and fire to understand and predict the complex environment of a fire, which can help in the evaluation of designs. Whether making choices by using published standards or by fire engineering techniques, there is a need to understand the progress of a fire in a building to make informed choices. Like any environmental design, it is necessary first to understand the conditions in which the construction must perform. The progress of a fire in an enclosed space follows a consistent pattern. This is shown in the typical fire curve illustrated in Figure 8.16.

For a fire to start there must be combustible material (fuel), oxygen and an ignition source. Fire in a building is really a series of fires starting at various time intervals and each following the basic development on the fire curve. Fire in each space requires ignition and then the amount and type of fuel and the rate of supply of oxygen will determine its progress. To understand how fire grows and spreads, it is important to remember that heat can be transferred by conduction, convection and radiation, and each will be involved to

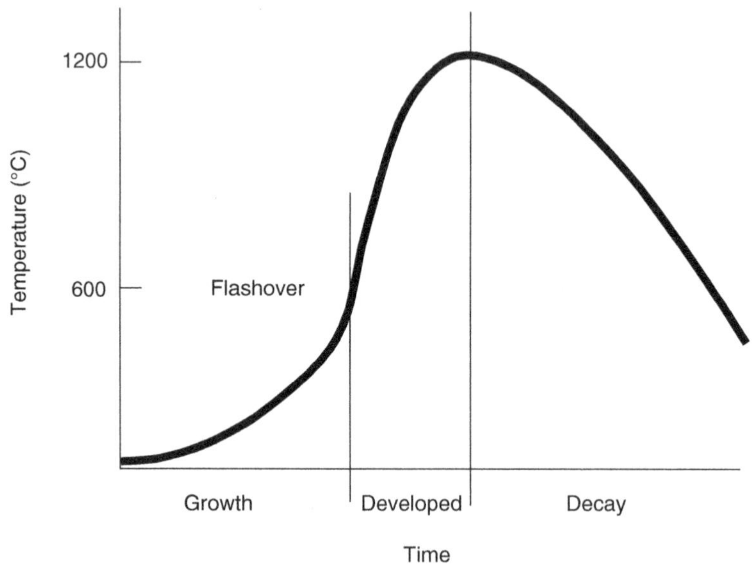

Figure 8.16 Typical fire time–temperature curve.

a greater or lesser extent depending on how well the fire is developed. These transfer processes are important in both the growth of the fire in each space and the generation of the ignition source for the spread to other parts of the building. It is also important to know that the objects and materials that combust give off not only smoke but also flammable volatiles, such as gases, which burn in the presence of oxygen, generating flames. This adds greatly to the heat evolution.

Fires normally start small. Only objects in contact with the original ignition source start to burn. Gases, heat and smoke will be given off. These will rise in convection currents, heating other possible fuel sources, which start to give off gases, and these all collect under the ceiling, where the surface area of this flammable material becomes much greater. If there is still oxygen to support the fire, when the flames reach the ceiling and the temperature is great enough the gases will ignite. The resulting increase in radiant heat will now cause ignition in most of the remaining flammable objects and materials (particularly surfaces or linings) in a very short space of time. This point occurs at around 600 °C and is known as flashover; the fire is then fully developed. If there is insufficient oxygen the gases will not ignite, but the objects that are on fire will continue to smoulder and give off heat. The introduction of oxygen as the door is opened will cause the sudden ignition of the gases, causing the phenomenon known as backdraft. The temperature in the space now rises rapidly and the potential for the fire to spread becomes much greater. Containing the fire at this point depends on the fire resistance of the elements such as walls, floors and the building structure. By following the development of the fire as described in this way it is possible to identify the technical solutions that could be used to make sure people escape and that the integrity of the building is maintained throughout the fire.

Fire prevention

The best defence against fire is to prevent ignition. Identifying possible sources of ignition and isolating them from a significant supply of fuel will accomplish this. If the activities or processes undertaken in the building produce heat, they should be carefully designed to be isolated some distance from any combustible material. These can often be easily identified as heat-producing appliances and flues. Minimum distance of timber components from flues is a good example of fire prevention. However, another major threat is the malfunctioning of any energy distribution service such as electricity and gas. Ignition from accidental sources such as discarded cigarettes is difficult to control by strategies involving the choice and detailing of the construction. However, treating fabrics with fire retardants has had major success in limiting the start of fires from accidental sources.

Fire detection and escape

If a fire does start, there should be a maximum chance of it being dealt with in the initial phase of the time–temperature curve. This involves early detection and fast extinguishing. These two functions are most often associated with the services provided in a building. Both detection and alarm can be either manual or automatic. The simplest means of detection and alarm that has been used is the provision of a bell (or triangle) to be rung by hand by the person who detects the fire. Alarm may still rely on an individual activating a building-wide alarm system automatically by breaking a glass disc in front of the alarm switch. Increasingly an automatic detection system will be specified using heat and smoke detectors to set off the alarms. Having detected the fire and raised the alarm, two activities must be considered together: the evacuation of the people and the provision for firefighting.

Given that the fire has been detected during the growth phase indicated on the time–temperature curve, heat will still be relatively low, although smoke may build up very quickly. If the fuel is chemically complex, the smoke may be acrid and/or toxic. Escape must be a priority. Most deaths in fires are from asphyxiation by hot smoke rather than burns due to the heat.

With very early detection it is possible to smother a small fire. Traditionally this would have been done with manually applied material such as sand or a fire-blanket, but nowadays a fire extinguisher would be used. There is a link here with the operational management of the building. The contents of all extinguishers are not the same. Products of water, foam, powders and CO_2 are available for different types of fire; clearly water should not be used on electrical fires or fat fires. If initial firefighting equipment is to be provided, then personnel must be trained in its use and the equipment must be checked regularly.

Automatic initial extinguishing can be achieved with sprinkler systems (or their gas equivalents if evacuation can be achieved first). These include a detection system that automatically releases the extinguishing medium, water in the case of sprinklers. In sprinklers each sprinkler head is its own detector (heat expands a liquid in a glass stopper which breaks and allows water to escape from that head only), so extinguishing can be as local or widespread as the fire. Even if the system does not extinguish the fire, it gains valuable time in that it considerably reduces the production of both heat and smoke, prolonging the initial phase on the time–temperature curve and the opportunities for escape. It does, however, require a large volume of water if the control is expected to be widespread as in large open warehouses. Here the public water supply may not be able to deliver at the required rate. Large storage tanks with pumps may have to be provided to maintain flow should the sprinklers ever have to be used.

As soon as detection and alarm have been established evacuation should commence. Again, the link between construction provision and occupation management is very clear. The design of the building must provide the means of escape, while it is the responsibility of the occupation management to ensure that the routes are known to building occupants and maintained unobstructed.

The design of the building for means of escape is initially a case of planning routes and determining their size to ensure a speedy and panic-free evacuation. Everybody should perceive that they have an equal chance of escape. Anybody in the space in which the fire starts must be able to move to areas that are on the escape routes directly linked to the evacuation point. Here they may meet people from an as yet unaffected part of the building also heading for the evacuation point. While most escape routes will utilise the everyday corridors and stairs, additional provision is often necessary because everybody would be using them at the same time. It is also necessary to provide alternative directions of escape so that individuals do not have to go through the fire to reach evacuation points. The distances and size of these routes have been determined empirically and are now provided in codes and regulations, and software modelling can also provide predictions of behaviour of fire and smoke spread, as well as people fleeing the fire in the early stages of evacuation.

All escape routes need protection from heat and smoke to allow safe escape evacuation. This protection will be provided primarily in the choice of materials and detailing of the construction that encases the route and ultimately in the stability of the structure of the building. To think about the properties of the material chosen and the appropriate detailing it is necessary to see how a fire spreads from its initial ignition to other parts of the building.

Building fabric fire performance

Fire development and fire spread are the two areas where the fabric of the building makes most of its contribution to fire design. The floor, wall and ceiling coverings, or linings, influence the development of the fire, and then fire resistance of the elements, both structural and non-structural, influences the spread. The role of the fabric can be identified as significant in the following two broad areas:

(1) Lining materials
(2) Fire-resistant elements
   - Internal wall and floors
   - Structural members
   - External walls and roofs

Linings are significant in the initial stages of the growth of a fire, within each space the fire-resistant elements are important as fire barriers and maintaining structural stability. Fire barriers also slow the spread of smoke within the building and, in some cases such as compartment walls and floors and in protected shafts, contain the fire within sections of the building for a specified period. In the design of party walls, external walls and roofs, the aim is to ensure that the fire does not spread to adjacent property. Structural integrity needs to be maintained throughout the fire, where the question of its potential performance in any rebuilding may be considered.

Fire performance of linings

The choice of linings can affect the time available for escape by slowing down the growth of the fire, but they may also contribute toxic gases and/or flammable volatiles, which may add to flashover gases or create hot droplets that may be a source of ignition. The properties of the materials for fire performance for the linings to the floors, walls and ceilings can now be identified. Combustibility, surface spread of flame, and fire propagation all indicate how quickly, if at all, the lining will assist the development of the fire. Smoke and droplet (burning particles) tests along with measurements of contribution to flashover gases give a complete picture to evaluate the choice of lining against fire performance criteria.

Fire resistance of elements

Having chosen materials for the linings, the next concern is the spread of smoke to escape routes and the fire igniting material outside the room. Controlling spread is the role of the fire-resistant elements. Performance is not just dependent on materials' properties but the design of the whole element, be it wall, floor, or structural components. This is recognised in that while we talk about lining materials, we also talk about the fire resistance of elements. Whole assemblies must be tested, including all the detailing, joints and fixings (and loading for structural elements), for a performance rating for fire resistance.

Fire-resistant elements may have the role of a barrier to smoke and fire and/or be a part of the building structure, the collapse of which would not only allow the spread of fire but also cause physical danger to occupants escaping and members of the rescue services.

There are three ways a fire-resistant element could fail and allow a fire to spread:

- Resistance to collapse (stability)
- Resistance to fire penetration (integrity)
- Resistance to the transfer of excessive heat (insulation)

The stability of a structural element may be affected if the fire reduces its load-carrying capacity to the point where the element collapses under the load it must carry at that stage in the fire. The integrity of a barrier may be lost if the element cracks, shatters or just burns away. This includes jointing materials that may burn or melt, or fixings that fail under the influence of the fire. The insulation of a barrier must be sufficient to keep the other side of the element from reaching temperatures that will cause ignition. More correctly, 'insulation' should be seen as heat diffusion. High thermal mass or endothermic reactions (chemical changes that absorb heat from the fire) will have a similar effect of maintaining the other side at temperatures below ignition for the duration of the fire, even though in thermal resistance terms the material of the element may be considered a poor insulator.

Internal walls and floors can be designed to contain fire and smoke, and hence control the spread of a fire within the building. Not all need the same degree of resistance to the passage of fire. The performance requirements will depend on two factors. First, the nature of the occupancy, including the anticipated maximum number of people in each area and their mobility. This will determine the actual time it will take to evacuate everybody to safety. Second is an estimate of the speed of growth and ultimate ferocity of the fire as it follows the pattern illustrated in the time–temperature curve (Figure 8.16). This will inform the design of building to contain fire and smoke in the event of a fire.

Another consideration that will determine the fire resistance of elements is the size and configuration of the building. To limit the fire resistance requirements for individual elements it is possible to design the whole building as several fire compartments, separated by compartment walls and floors designed to contain the fire completely for a specified period. While the specification of these elements may be particularly onerous, the elements themselves do reduce the demands on the other elements inside the compartment as the risk to both people and property is significantly reduced.

Openings are potential breaches in the integrity or insulation of the element. Two types of opening can be identified: doors (and opening windows) and spaces for services to pass from one area to another. Doors must contain both heat and smoke for a prescribed period. Not only must the door have properties of integrity and insulation but also the gap around the door is particularly vulnerable in spreading the fire to the outer surface of the door. The dimension of the closing edges and the coverage offered by the rebate become a particular aspect of detailing. Closing edges on double doors may have to be fitted with intumescent strips. These materials expand on heating, sealing gaps and ensuring integrity and offering an insulating layer against heat transfer.

The small spaces created around the services passing through walls and floors need to be filled or stopped with a fire-resistant material to maintain the integrity of the element. The services are often carried in voids either above the ceiling or below the floor.

The convenience of these voids is that the services may pass unhindered above or below the rooms independent of the partitions between rooms. In fire design these are known as cavities and have the potential to spread smoke and fire rapidly to other parts of the building. There is a need to provide barriers in these cavities to prevent the spread of both smoke and fire. These cavities should not extend across compartment walls or through compartment floors.

Another danger is spread via vertical shafts provided in buildings for circulation (stairs and lifts) and for services. Once breached by fire, they will act as flues and transmit heat, smoke and volatile gases to higher floors very quickly. The fire resistance of these shafts becomes an important aspect of the fire design. Protected shafts between compartments must have walls with high resistance ratings and the openings such as doors and service entry points must be fully protected. Since the stairs are usually needed for escape routes, preventing the intrusion of smoke will be a major performance requirement, possibly by the introduction of positive air pressure. Cavities within the building, for example cavities within walls and cladding systems also provide a pathway for the rapid spread of smoke and fire, especially to higher floors. To prevent the spread of fire and smoke the cavities must be designed and constructed with cavity barriers.

As well as establishing barriers to control the fire and smoke spread it is necessary to ensure that the structure of a building does not collapse or damage the barriers for the duration of the fire. There may also be some consideration of the integrity of the structure at the end of the fire for its use in any rebuilding work.

Structural materials in fire

Elevated temperatures change the properties of many materials, including strength and stability of load-carrying elements may be lost without any apparent physical fire damage to the element. Structural sections exposed to fire may not burn, but the rise in temperature may cause expansion and/or a significant loss of strength. The section may deflect and even collapse, as it is no longer strong enough to carry the loads it was designed to carry.

The mechanisms are different for the major structural materials. Steel has a high thermal conductivity and low thermal capacity, which leads to a rapid rise in the temperature when exposed to fire. There is a rapid loss of strength as steel has lost something like 50% of its yield strength at 600 °C, well within the temperatures reached in a building fire. Steel can be fire engineered but is usually protected from the heat. Casing or spraying the steel with either an insulating material or a material that can absorb the heat can achieve this. Absorbing can be achieved by either high thermal capacity or endothermic reactions, where chemical changes in the materials use the heat of the fire. These casings and spray-applied materials must remain attached and not crack during the fire. Other strategies that have been used include filling the structure with water to absorb the heat or putting the structure on the outside of the building, but these options have not gained widespread application.

Reinforced concrete has a high thermal capacity so, unless the sections are thin, the rise in heat during a fire is slow other than near the surface. If the surface cracks or spalls, or if the steel reinforcement that is near the surface is allowed to gain heat, the section may fail. This focuses attention on the concrete providing cover to the reinforcement. The aggregates in the concrete and the depth of cover are the specification items of significance to the fire resistance of reinforced concrete.

Timber has a completely different mechanism associated with achieving fire performance. Timber burns, but the charred products of combustion remain attached to the section and insulate the unburnt timber from oxygen and heat. Burning slows considerably as the fire progresses. If unprotected structural timber is specified oversize to account for a charred layer, the timber will carry the load for the duration of the fire, but it will need to be replaced in any rebuilding that follows the fire.

Fire performance of external elements

Having specified internal barriers and the structure so that they perform during a fire, there remains one other major threat. If the fire breaches the external envelope, it not only will gain a source of oxygen but also may spread between adjacent properties. There are two ways that this spread between properties could happen. The first is spread along external surfaces to adjacent buildings, particularly along roof surfaces. The second is directly from areas of the building breached by the fire, such as windows. If the space between the buildings is large enough, the risk of spread is eliminated. However, in most cases there is a need to have some degree of fire resistance in the materials used for the walls and roof. The closer the buildings are together, the greater the area that must be fire resistant. Codes and regulations give guidance on how much unprotected area can be allowed against notional boundaries between buildings.

Specification of fire resistance

These performance requirements need to be specified in terms of time. There remains the problem that while evacuation can be established in real time the actual time an element will hold back the fire depends on the ferocity of the fire. For comparable performance between elements, testing must use a standard fire, one with a predetermined time–temperature curve. Hence the times quoted for fire resistance are not real time but standard test times. The required fire resistance must be chosen by considering both the required evacuation real time and the ferocity of the fire defined by its time–temperature curve.

Firefighting provision

There remains one other area of consideration for the design of the building, that of the provision for firefighting by the fire service. Earlier discussion covered the provision of equipment for fighting the fire by the occupants early in its development. Once a fire starts to develop this equipment has limited value. It becomes necessary to give access and provide services such as hose reels, wet and dry risers, and foam inlets for the use of the fire brigade. This will require consideration to be given to the access of fire appliances into the grounds of the building and, possibly, on large buildings, even specialised personnel access in the provision, for example, of fire lifts. Like so much fire provision, this is a matter for initial design and fire engineering, but it also must be managed 24 hours a day throughout the life of the building to ensure fire safety.

Threats from within society

Some threats come from the society within which the building is located. Like the natural threats (acts of God) and fire, they are threats to both person and property. In most politically stable societies direct threats to the person from strangers are relatively low unless associated with a threat to property. These threats are most usually associated with crime, possibly including terrorism. They are characterised by the means individuals are willing to use to act out the threat. These are complex sociological challenges, and it has been suggested that the quality of the built environment can play its part in reducing the threat as well as providing defence against it. The nature of the threat is highly dependent on the activity to be housed in the building. For the most part it is concerned with the intrusion of an external threat, but there are situations where the threat is being housed (e.g. prisons) and the risk is to the activities being carried on outside the building.

Whichever of these is the case, the approach to design is likely to follow that identified for fire: deterrent or prevention, followed by detection and alarm, followed by the provision of resistance to the forces of the threat to limit injury to persons and damage to property. Unlike fire, however, the perpetrator of the threat has reasoning and therefore defining any set pattern of attack is difficult. For instance, there is for each perpetrator a fear of detection and an assessment of the risk of being caught. Assessing the threat involves assessing determination and the means the perpetrator is willing to use.

Personal accident

There is the need to reduce the risk of personal accident while using the building under normal operational conditions. Evaluation of the potential for accidents in the building is the same as the approach taken to health and safety on construction sites. It starts with the identification of hazards and an assessment of the potential to do harm. When the chance of an incident causing harm is established, the risk can be assigned. In the daily operation of most buildings safety equipment is not available or appropriate, so the hazards should be eliminated in the design. This determines much of the dimensioning and detailing of ramps, staircases, balustrades and handrails. Safety issues are implicit in specifying fixings for fittings where failure during use would cause harm. Another example is specifying floor finishes that are not likely to become slippery when wet. In addition to reducing the risk of personal accident, materials should not be chosen that might be injurious to long-term health.

8.8 The private environment

The origins of the requirements for privacy are the conventions of society. Unlike safety, the needs do not arise from the existence of a threat but from a sense of what is right and wrong, and what is perceived as acceptable or unacceptable social behaviour. To understand the needs of privacy it is necessary to understand the ways of the society and the beliefs of its people. This can be challenging in a multi-cultural and multi-faith society and requires consideration at the early stages of any project. The concept of 'family' dictates much about the form of domestic dwellings while social norms and organisational culture will determine the form of commercial buildings. This is also linked to the concern for image and status, which may influence appearance and the division of internal spaces.

The feeling of privacy comes not from the absence of the sense of impending physical danger but from the proximity of others invading one's personal space, even if no physical barriers exist in a room. Social convention often requires physical separation, involving limiting the attention of others seeing or hearing or in any way being aware of the activity associated with the use of that space. Thus the major requirements for privacy are acoustic and/or visual separation. The need for privacy adds to the performance requirements of the enclosing elements of the building. Although predominantly walls, it includes the construction of floors, ceilings and service distribution routes. The analysis and evaluation of the performance of visual separation needs no complex theory or testing other than that of direct observation. Predicting acoustic performance on privacy is less direct and has been considered earlier in this chapter.

Physical Behaviour Under Load: Chapter 9 AT A GLANCE

What? To consider structural behaviour it is first necessary to understand the external loading conditions, primarily dead, imposed and wind loading, although they may include impact and moving loads. These loads will move or disturb the structure unless it is stabilised, after which the load will distort and potentially destroy (break) the members. All parts of construction are subject to at least some of these loads and therefore they must be considered for strength, damage resistance and adequate methods of fixing. There is for all buildings a recognisable structural system transferring these loads to the foundations. This system will comprise a series of structural members and supports jointed together to provide a load path through the building.

Why? The way members are connected affects the distribution of internal forces. Pin joints, which allow rotation, do not transfer bending, while fixed or rigid joints do redistribute bending stresses into the support. Five basic structural members can be identified: the strut, tie, beam, wall and slab. Each can be used for a range of loading conditions and support arrangements, which will determine the distorted shape and will give clues to the distribution of internal forces that will be generated. These forces are compression, tension, bending, shear and torsion. Closer analysis of these forces shows them all to resolve into either compression or tension. In addition to the distorted shape that arises in the direction of loads, lateral distortions will appear in slender members in compression at right angles to the direction of loads. Eccentric loads will also introduce new distortions, leading to new internal forces.

When? Analysis of loading will be undertaken by engineers early in the design process, which may influence the general layout of spaces and massing. As the design develops from generic forms into more specific and detailed design, the structural analysis will inform the sizing and spacing of structural members and the manner in which the various components must be joined. This informs the choice of the structural system, for example loadbearing masonry or a framed construction. It also informs the type of materials to be used, for example brick, timber, steel or concrete.

How? When members are arranged into building structures it is possible to see the way in which loads are transferred to the foundations. With the whole structure stabilised, and the pattern of external and internal forces established, the shape, size and materials for each member can be specified. The shape and size of the cross section of the member will determine the area over which the internal forces will act. This will give the stress that will be experienced by the material. Knowledge of the stress and strain characteristics of the material can be used to check the suggested shape and size for sufficient resistance to keep suspected failure within factors of safety.

9 Physical Behaviour Under Load

All parts of construction are subject to loads, even if it is only their self-weight, and there is a need for a building to have a system of structural members joined and jointed together to transmit the loads to the foundations. It is necessary to determine the way the loads are applied and how the support is provided for each structural member, together with the way they are jointed to one another. This will establish the pattern of internal forces that will generate the stresses in the materials chosen. It is suggested that visualising the distorted shape of the structural members will indicate the pattern of internal forces. The system must contain sufficient members and be jointed in such a way as to remain stable, particularly against horizontal forces such as wind loads. With knowledge of the stress–strain relationship of the material this pattern of internal forces gives the clues to the efficient shape of the structural members and the economic size.

9.1 Forces - external and internal

As soon as any part of a building is created it will be subject to forces even if only from its own weight, and therefore it must have strength and be adequate fixed. There are some elements of the building whose function would be recognised as part of its structure. They maintain the shape of the building and transfer loads from one part to the next to reach an existing stable support, normally the ground, which must take the accumulated loads and therefore becomes part of the structural system. These external forces, or loads, are predominantly generated by the weight of the construction and the objects and people inside the building. However, there are also environmental loads, such as wind and snow, which create forces on parts of the building exposed to the elements that must be transferred successfully to the ground.

External loads create internal forces within the structural elements. The nature of these forces and the mechanism by which resistance can be provided become the focus for the analysis of the physical behaviour under load. These elements of structure must ensure the building remains stable, has limited distortion and does not break. Instability, distortion and breaking are the possible failures that will have to be identified, and a level of performance will have to be established for each.

The initial design task is to identify a system of structural members and supports that, when jointed together, provide a stable structure against the loads that will be applied to them. Loads applied to one member in the system must be transferred to a support

Barry's Construction Technology: Analysis and Choice, Third Edition. Stephen Emmitt and Tony Bryan.

through some form of joint. The load being transferred through the joint effectively places a load on the support that becomes the next member in the system. This load path must be traced through the structure. Each member and joint must be capable of resisting the accumulation of loads being transferred through the structural system. It is important at this initial stage to ensure that the structural system has all the members necessary to ensure its overall stability. The initial effect of the loads will be to displace or rotate the members to which they are applied. This can be resisted at the joints between structural members, but often no such resistance exists. It is not until the additional stability members have been identified that the overall structural system has been devised.

Once the structure has sufficient joints and members to achieve overall stability, the loads will appear to distort the structure. This distortion arises because the externally applied loads induce forces inside the members and joints that stress the materials of which the structure is made. When materials are stressed, they undergo dimensional changes, known as strains, that cause the members to distort. As stress increases in the material it will ultimately cause the member to break. Understanding structural behaviour and undertaking the design of structures is concerned with establishing stable structures against the external forces created by the loads, and the analysis of the internal forces and the stresses the loads induce, along with the strains that arise causing the distortions. The loads define the conditions under which structures must perform. Failure or loss of performance can occur if the loads:

- disturb the structure, making it unstable
- distort the structure too far so it cracks finishes, transfers loads to non-loadbearing members or looks and/or feels unsafe
- destroy part of the structure by breaking members or connections at the joints.

The proposal or suggestion is for an arrangement of structural members with choice of materials and approximate sizes to support the loads. An evaluation of the response of the proposed members to the loads is by analysis of the distortions and the distribution and size of the forces within the structural members to see if they will maintain the loads without collapse. It is necessary to look at each of these stages in more detail.

Loads and loading patterns

Loads on the building come from the weight of people and the objects in the building, the weight of the building and from the environmental conditions to which the building will be subjected. What they have in common is that they all apply forces to the construction, which must be resisted if the building is to continue to remain stable and then not distort excessively or even break. The most usual way to classify loads is as follows:

- Dead loads – the weight of the construction
- Imposed loads – from the activity within the building and environmental loadings from snow
- Wind loads – an environmental loading from the wind

Before the characteristics of each of these types of loads can be appreciated it is necessary to identify what needs to be known about a load before it can be evaluated as a force on the building.

The following represents a definition of a load:

- ❑ Magnitude or size
- ❑ Distribution
- ❑ Direction

Magnitude is the total weight or force to be taken by the structure, while distribution identifies the pattern on the member. Distributions are often seen as a single point, uniformly distributed, or possibly moving and impact loads. Direction completes the specification. Most loads on a building act vertically downwards under the influence of gravity; some will act at other angles (thrust at the bearing of an arch), with some, such as the wind, even acting vertically upwards under certain conditions.

It is now possible to see the basis of the classification. Dead loads are permanent loads where their size, distribution and direction can be established with some certainty. Imposed loads are variable so can be identified with less certainty. While magnitude can be established within certain limits, the patterns of loading need to be established to identify design loading conditions. There may be some objects in the building, such as machinery or storage arrangements, that are so much larger and more permanent than any of the other components of the imposed load that they would be treated separately, with their position defined with the characteristics of a dead load. However, most imposed loads can be brought together as static uniformly distributed load defined by the notional use of the building (e.g. domestic or office).

Wind loading is also variable. Its inherently dynamic nature coupled with its variability in intensity and direction make a definition of size, distribution and even direction problematic. As wind speeds up as it flows over and around building it creates low-pressure areas, resulting in uplift forces, particularly across roofs. If vortices are created at edges, even greater suction forces can occur. To provide a definition of the wind load (and hence the forces on the construction) that can be used in a structural analysis, certain simplifications have to be made to obtain an equivalent static load. These simplifications must, however, ensure that the effect predicted is at least as demanding as the effect that will be experienced by the real structure under the complexity of the real environment. Procedures for this are now well established and published in codes and standards.

Another condition that must be allowed for is the possibility of accidental loading causing a local structural failure that leads to a more general failure of a major part of the building. This can lead to a disproportionate collapse in that the final proportion of the building affected is disproportionate to the local nature of the original failure. Some buildings will have to be considered for other accidental loading, such as impact from aircraft for high-rise buildings and blast for security buildings or those deemed to be at high risk from terrorist attack. The natural world can also create loadings from ground water or earthquake. This serves to highlight the importance in performance specification of defining the conditions that should be designed for, either as the limits considered as normal or in terms of extreme yet potentially disastrous conditions. However, most buildings are only designed for dead, imposed and wind loads plus measures against disproportionate collapse. Designing for extreme loading in most buildings would not be seen to have value in the inevitable extra building cost if the probability of them occurring would be very low.

9.2 Basic structural members

A limited number of types of structural members can be employed in buildings. They act either to span around spaces or to provide support or stability to hold the spanning elements in place to generate the spaces within the building. While the common names for these structural elements – walls, columns, beams and slabs – give some clues to the position in the building and their basic form, understanding their structural behaviour needs closer analysis:

- Where the loads are applied
- What support is provided (to make stable and stop disturbance)
- How the stable member distorts

By visualising the distorted shape it is possible to identify the distribution of internal forces that would break the member. Figure 9.1 illustrates perhaps the most straightforward structural members: the strut and the tie. The strut (commonly identified as a column) is loaded and supported in such a way that it experiences compression, identified from visualising the shortening caused by the loads. By contrast, the tie is loaded and supported so it goes into tension, indicated by visualising lengthening.

Figure 9.2 illustrates the beam, a spanning member where the loading and support pattern causes the member to bend. Two beams are shown in Figure 9.2, each with their distorted shape, indicating that where the material in the beam shortens it experiences compression and where it is stretched it experiences tension. The maximum lengthening

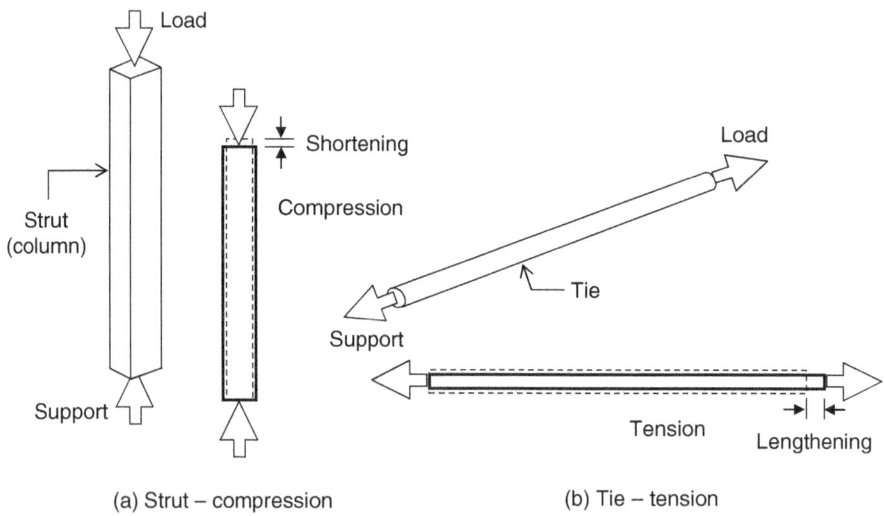

Figure 9.1 Struts and ties.

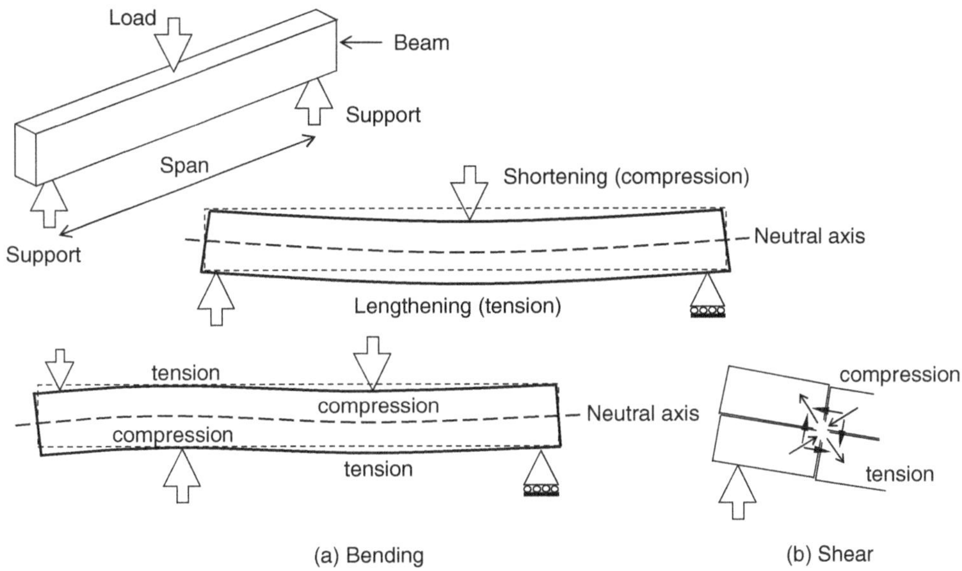

Figure 9.2 Beams.

and shortening will occur in the top and bottom of the beam, with no distortion being experienced along the neutral axis. All bending members have this pair of tension and compression forces acting together across the depth of the section. All bending members have a second pair of tension and compression forces, known as shear. This time the forces are greatest at the neutral axis of the beam where they act diagonally across the material of the member.

Struts, ties and beams have direct compression, tension and bending forces respectively. The other two major structural members, walls and slabs, can have different load and support patterns giving different sets of internal forces. Figure 9.3 shows how a wall can be in compression when taking loads through the building, and it may be in bending if resisting horizontal forces such as wind on the façade or ground pressure as part of a basement structure. The shear-loading pattern is created when a wall is used to stabilise the racking effect of wind forces on building structures, as discussed later in this chapter.

Similarly, slabs can be subject to different internal forces depending on the loading and support arrangements. Figure 9.4 shows that this is mainly a distinction between suspended and ground-supported slabs. Suspended slabs will be in bending. The illustration shows support on only two edges, creating a one-way span, but it is possible to support the slab on all four edges, creating two-way spanning, to reduce sections or offer potentially greater spans. Ground-supported slabs carrying uniformly distributed loads will experience compression from the applied loads. When slabs are placed under columns and walls as foundation, the forces they experience depend on the ratio of the depth to the width of the foundation. If the foundations are comparatively wide, the foundation must act in bending for the load from the column or wall to be transferred evenly to the ground. When the slab becomes narrow, the predominant force experienced by the foundation is a punching shear.

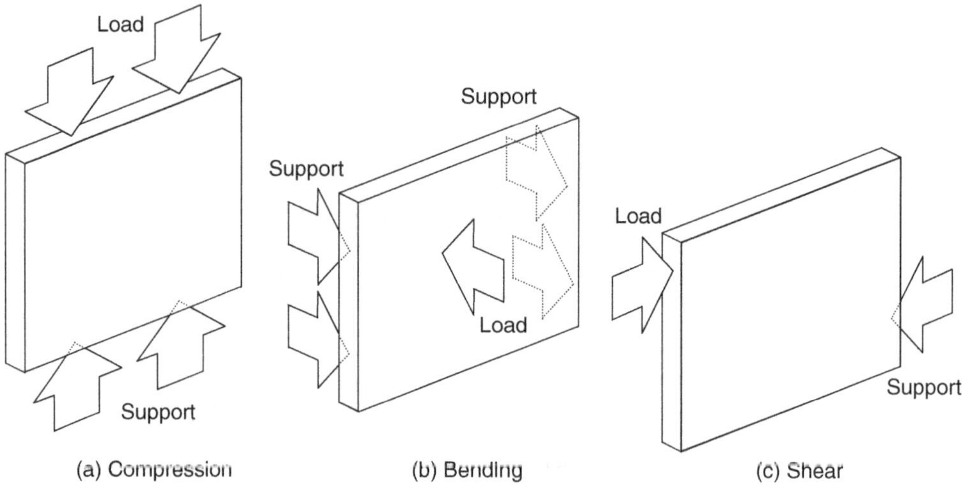

Figure 9.3 Walls.

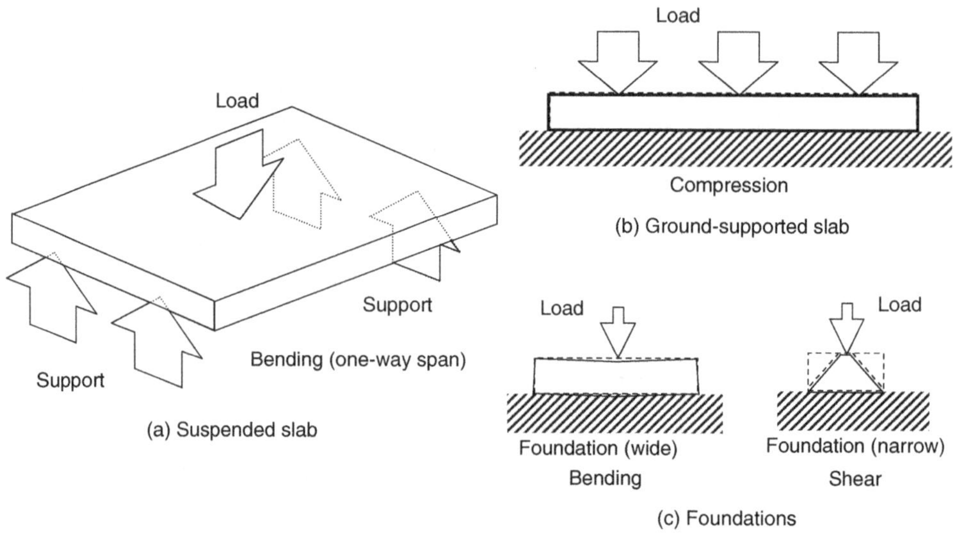

Figure 9.4 Slabs.

Lateral distortions

Figures 9.1 and 9.2 show the shortening of the strut, lengthening of the tie and deflection of the beam as distortions that arise as a direct result of the application of the load and in the direction of the load. These are not the only distortions that occur. Lateral distortions may be experienced in slender members where compressive forces are experienced. These are known as lateral distortions because the member bends in a direction at right angles to the direction of load. Slenderness, as the name suggests, is primarily a function of the height

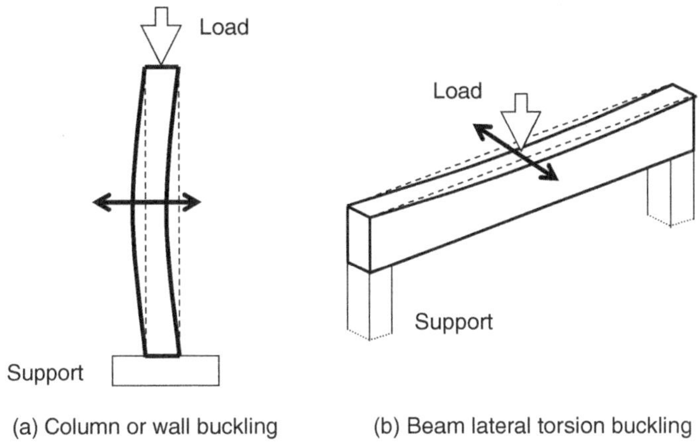

Figure 9.5 Lateral instabilities.

(or length) to width ratio, but it is also dependent on the way the member is connected to the building structure. The effect of the connection will be explained later.

Observation of the internal forces that develop in the members discussed in Figures 9.1 to 9.4 show that compression zones develop in all types of members except ties. However, not all these members become slender in a direction at right angles to the load. Again, by observation, perhaps the most obvious potentially slender compression member is the strut and the potentially slender compressive zones that arise in beams. The lateral distortions, known as buckling, which can occur in struts (column or wall section) and beams, are shown in Figure 9.5. In columns the buckling induces bending as the whole section is in compression. However, in beams the compression zone buckles while the tensile zone remains straight. This induces a twisting action known as torsion. The whole lateral distortion effect is known as lateral torsional buckling.

Less obvious is the possible buckling associated with shear. In thin web sections in beams the principal shear compression force shown in Figure 9.2 may develop buckling. In thin shear walls, shown in Figure 9.3, the buckle will occur across the diagonal, giving a potential bending failure like that of the buckling strut. It is important to realise that the buckling distortion does not replace the distortions of the direct loading, but when it occurs the structural member experiences both. The two sets of internal forces develop together. For instance, in a column already experiencing compressive forces the buckling adds bending, increasing compression on one side but reducing the compression where tensile bending forces from the buckling develop.

Eccentric loads

In identifying all the distorted shapes and internal forces so far it has been assumed that the loads have been applied, and support has been provided at the centre line (on the axis) of the member. These are known as axial loads, acting through the centre of struts and ties and along the centre lines of beams and walls. Although it is desirable for the economy and efficiency of structural members to be axially loaded, it is not always possible. Loads that act off the axis are known as eccentric loads, eccentricity being a measure of how far the load acts away from the axis.

The application of eccentric loads may cause the member to overturn if the dimensions and/or connections are not sufficient for it to remain stable. However, assuming the member is not disturbed in this way, eccentricity creates a further distorted shape and therefore changes the distribution of internal forces. In columns and walls they induce bending. In beams if the ends are restrained from rotation the effect is torsion, like lateral torsion buckling.

Like lateral distortions the additional distortions caused by eccentricity do not replace the direct loading effects but add to them. The effect in the column is, therefore, as described for buckling. Already in compression, the bending forces will increase compression on one side and the bending tensile forces will relieve some of the compressive force on the other. Because eccentricity induces a bending effect, this is likely to be the direction in which buckling occurs, given that the member has the same slenderness in all directions.

9.3 Curving (and folding) structural members

The members discussed so far have all been straight or flat. Curving the member may increase the manufacturing and assembly cost, but it does provide structural advantages and may therefore be worth considering. The structural advantage is associated with spanning members. To develop sufficient resistance to span in bending, the material in compression and tension from the direct bending effect (i.e. not the shear) must have sufficient separation to develop the moment of resistance. This makes beams and slabs deep if the spans become too large. This increases the depth of the structural members and, unless some of the solid section can be removed, increases the self-weight, adding to the dead load to be carried by the beam or slab.

Advantages can be gained from curving both in the direction of the span and at right angles to the span, but for different reasons. Consider the members in Figure 9.6. These

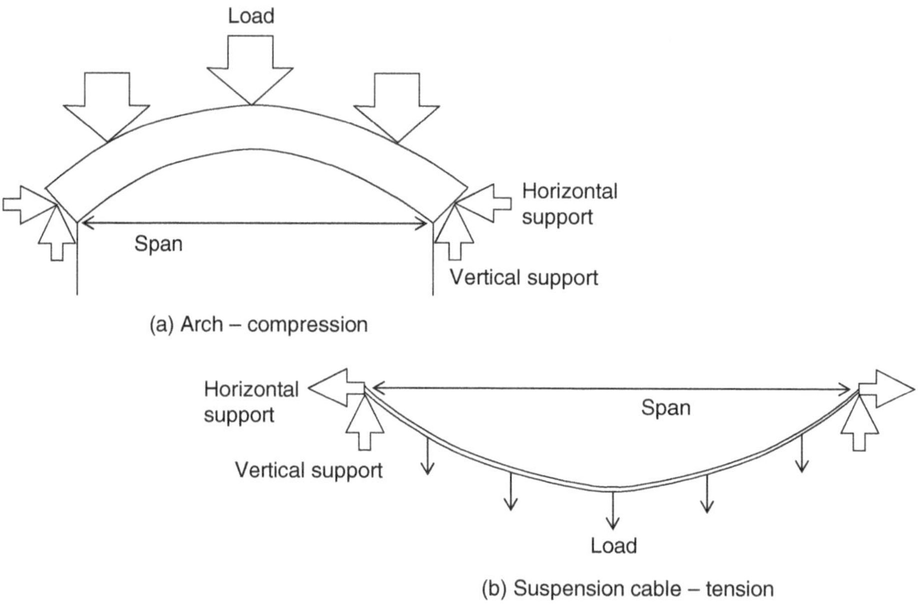

Figure 9.6 Curved structures across the span.

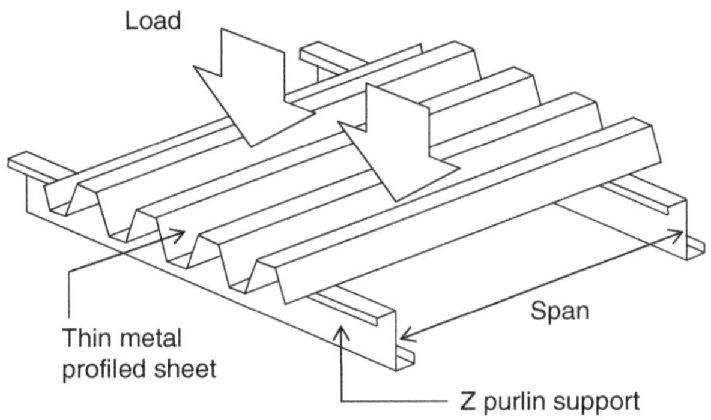

Figure 9.7 Curved (folded) in the direction of the span.

members that curve across the span are known as arches and suspension cables. Two significant aspects should be noted. First, they require both a horizontal as well as a vertical support and, second, the internal force becomes pure compression and tension respectively: they are effectively curved struts and ties. This allows materials and/or components naturally strong in either compression or tension, but not both, to be used in the construction. Less material is needed, and more material options are available. When slabs are curved like arches, they are known as shells, and when allowed to hang suspended they are known as nets or fabric structures (tents).

Curving and folding at right angles to the span gains a different advantage. It gives the depth that is needed to act in bending from thin sections. The most common use of this technique is for roof sheets or decking, as shown in Figure 9.7. Thin sheet materials can span many times that of the flat sheet form. Again, less material is used, but there is a manufacturing cost associated with the need to form and control the folding or curving. Curving in this direction is still a bending action and therefore the supports are only required to take vertical forces.

9.4 Structural connections

In the analysis of the five basic members (Figures 9.1 to 9.4) there is an implicit assumption about the way the member is connected to the support. All the distorted shapes assume that the member can rotate if it bends and can shorten or lengthen freely in response to the distorting effect of the loads. Whether this can happen depends on how the connections are detailed. There are advantages to restricting the movements at the connections, particularly that of rotation, to induce different distorted shapes to redistribute the internal forces to reduce the size of the member. This does sometimes increase the manufacturing and assembly effort and therefore has a cost to be set against the saving in section size.

To think about the effect of restricting movement at the connection on the potential stability and distorted shape of the member requires consideration of three 'pure' actions, shown in Figure 9.8. They are shown diagrammatically as they are idealised structural joints. The detailing to gain pure action is often difficult and expensive. The roller joint allows both rotation and freedom to move against the shortening (or lengthening) effect. Pin joints allow rotation, and rigid (fixed or moment) joints allow neither.

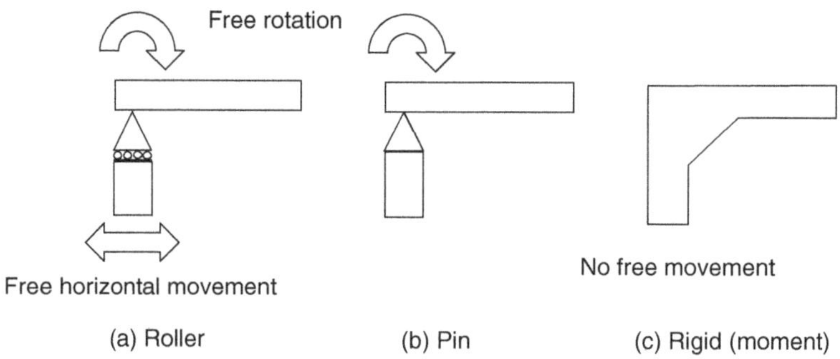

Figure 9.8 Structural action of connections.

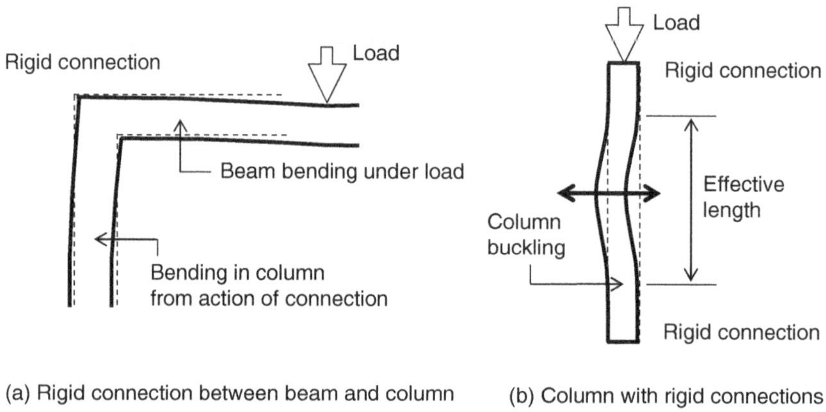

Figure 9.9 Effect of rigid connections.

Figure 9.9 shows the effect of making the joints rigid on two of the conditions already considered. The first is making the joints rigid at the end of the beam. The support now has a rotation (or moment) action to resist. If the support were a column with no balancing force from the other side, for instance from a similar beam with similar loading and similarly fixed, the column would go into bending. However, the internal forces in the beam are redistributed along the beam and the maximum force in the beam is reduced, but bending forces appear at the supports where there were none when the joint was assumed to be a pin. Tensile forces appear at the top of the beam at the support, the balancing compressive force being at the bottom. This redistribution of the bending along the length of the beam reducing its size is the basis of the portal frame, which can achieve long spans economically where a simply supported beam could not.

The example in Figure 9.9 is the effect on buckling in a column where the connections are fixed. Because of the moment resistance experienced at the end of the column, the effective length of the column is reduced and hence the slenderness is reduced. This column will take greater loads before it buckles.

Most connections are detailed for ease of manufacture and assembly rather than to ensure a pure structural behaviour. They give these freedoms of movement to a greater or lesser extent and therefore the real instabilities and distortions may not be exactly those assumed, and different internal forces will develop. However, most simple connections behave near enough to a pure form (normally pinned) for the assumption to be safe to use with the practical detailing. If there is doubt about the degree of fixity (often known as semi-rigid joints), an estimate of the effect must be established so that a more realistic distribution of internal forces in the members can be calculated. If the structural advantage of making the pure behaviour outweighs the additional cost of manufacture and assembly, details must be devised to ensure the pure behaviour is achieved.

9.5 Grid members – pin-jointed frames

All the structural members considered so far, flat or curved, have been solid. Curving across the span was worth considering for bending members to generate pure forces (compression or tension) and reduce the dead loading that occurs as beams and slabs are designed to span greater distances. Pin-jointed frames are another way to achieve the same ends. This is a very versatile idea, and many forms of these frames are used, both as plane frames (having little width, like beams) and acting in three dimensions like slabs, both flat (double-layer grids) and curved (single-layer grids as domes) structures.

The truss is one of the simplest forms of the pin-jointed frame and will serve to demonstrate the principles. It is shown in Figure 9.10. This is a spanning member with loads and supports shown and all pinned joints are assumed. Taking each member of the truss one by one, it is possible to trace the load and support of each to determine its distorted shape. The load applied to node a is directly transferred into member 1, which gains its support from node b. This will shorten member 1, putting it into compression. It must be designed as a strut (note the load will be axial but it may buckle). At node b, forces are transferred into member 2. Consider the effect on member 2. A downward force is experienced, which is supported at node c. Member 2 is stretched: it is in tension and therefore acts as a tie (note not subject to buckling). Forces at node c now pull down on members 3 and 5, pulling them together, rather like the action of an arch. This puts compressive forces into these members that are continued through members 4 and 6 to nodes e and g respectively. This generates

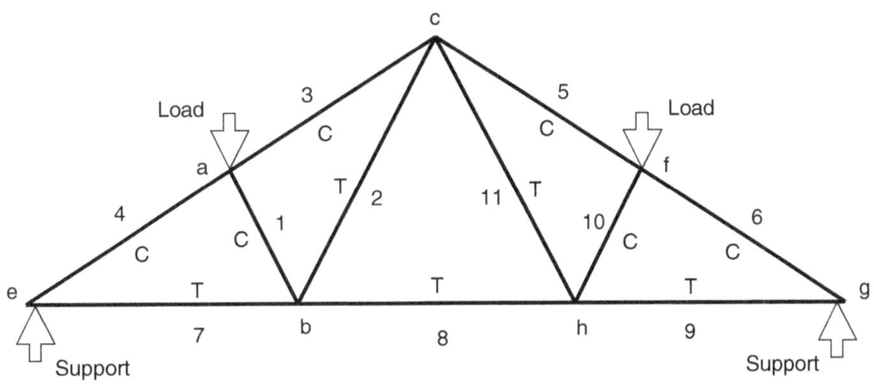

Figure 9.10 Pin-jointed frames – the truss.

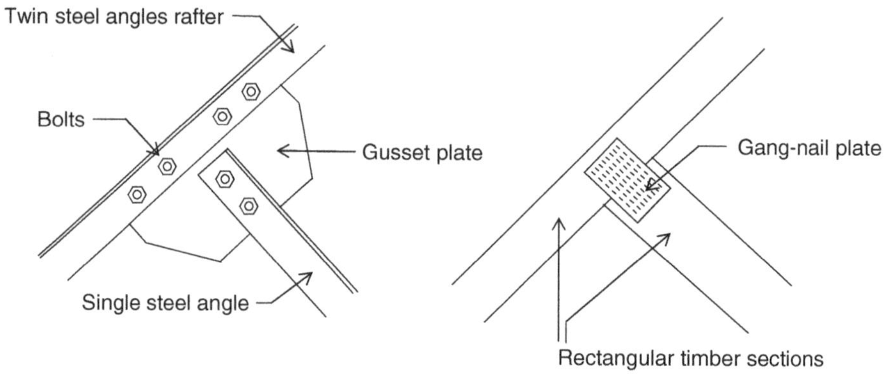

(a) Traditional steel truss

(b) Timber trussed rafter

Figure 9.11 Detail of joints in pin-jointed truss.

a need to support a horizontal force at e and g, and members 7, 8 and 9 give mutual support to e and g. These all become ties. By the same logic, member 10 is a strut and 11 a tie. All the forces are resolved within the frame as either struts or ties. The supports now only experience vertical loads, assuming the connection of the truss to the support to be a pin.

This analysis is based on the assumption of the truss acting as a pin-jointed frame. There is, however, a need for the connections to be practical while being able to assume that the structural action is that of a pin joint. Figure 9.11 shows a typical detail in steel and in timber of the connections at node a. The illustration shows that members 3 and 4 are all one piece, but in this case the relatively low stiffness of the members themselves means that the assumption of pin jointing is safe.

9.6 Building structures – wind stability

When considering structural members, so far, the loads have all been vertical, essentially from the dead and imposed loads on the building. Actions of forces other than vertical have been identified, but these have come when members transfer their loads to other members as the loads find their way to the ground. When visualising a whole building structure capable of carrying these vertical loads, it may not be stable against wind loads. Only when the whole building structure is identified can the effect of wind loads be considered.

Wind loads can act vertically. One of the most destructive aspects of wind to the fabric of the building is the negative, uplift forces on roof structures as the wind speeds up as it is forced to go round and over the building. Where roofs are over large open spaces, the wind can get into the building and create positive pressures under the roof, increasing the effect of the uplift forces. However, the major initial concern to the overall structure safety is the horizontal forces acting on the sides of the building. These can be both positive and negative forces, often being positive on one side and negative on the other, having a cumulative effect on the structure. One further aspect of the wind is that, while it may have a prevailing direction, it will sometimes blow from other directions and therefore the building must be designed to withstand wind from all directions. Wind or overall stability of a building is a major design criterion. Additional members may be necessary to stop wind disturbing the structure. These members and/or connections allow the forces to be distributed through the structure and safely transmitted to the ground.

The major structural systems used for buildings are loadbearing walls and framed structures, which also include most of the long-span roof solutions. The predominant instabilities and the measures to overcome them vary between these two structural systems. However, there are only a limited number of strategies to overcome these overall instabilities, which can be used alone or in combinations:

- Increasing dimensions (against overturning)
- Changing joint type (from pinned to fixed)
- Providing additional support members (buttressing, bracing, etc.)
- Changing shape (curving, folding, etc.)

Some of the alternatives used for walls and frames are shown in Figure 9.12. The major instability failure for walls is overturning. Widening and embedding are options most likely to be used in retaining walls resisting the horizontal forces from ground pressure rather than from wind. They have limited application in the superstructure against wind. Walls above ground are almost certainly likely to be slender and will rely on return walls and piers (buttressing) plus the stiffening effect of floors.

In frames the instability is racking. The frame can rely on moment (fixed) joints between the structural members, but this often is an expensive option. For frames with pin joints, bracing can be used, although it is possible to utilise other parts of the construction, such as

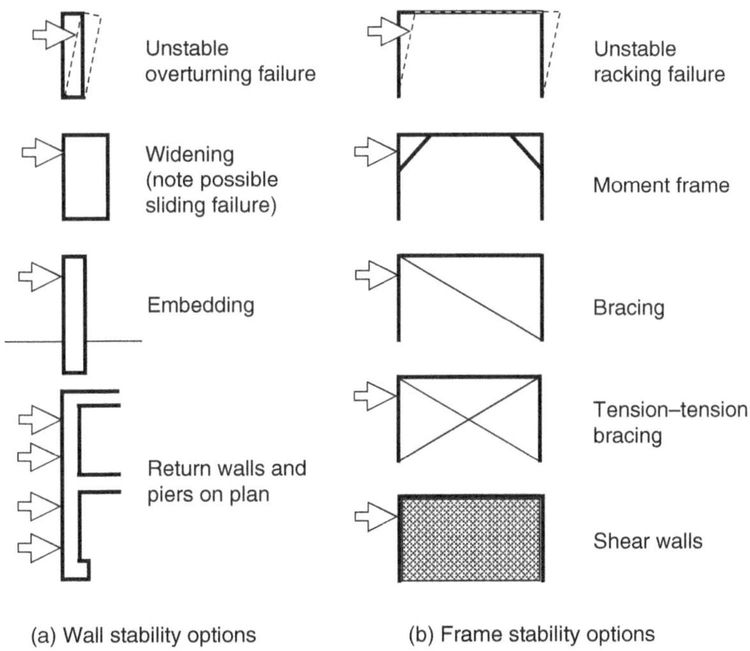

Figure 9.12 Stability of walls and frames.

internal walls or lift shafts, as shear walls (core stability) or the external hull either braced or with moment resistance from the structure and/or façade. It is not necessary to stabilise every frame. If a sufficient vertical section of frames is stabilised the whole building will be stable. The number and position of these stabilised vertical sections will vary, but structures must contain at least one section to stabilise each direction and at least one other section to resist a rotational collapse of the whole building. An example of this is shown in Figure 9.15.

Whatever additional support members are used these then become members with their own connections, distortions and internal forces. Consider the single diagonal bracing shown in Figure 9.12. Depending on wind direction, it may be either a strut or a tie. It must be designed as a strut as it is almost certainly slender, and this makes it susceptible to buckling. The benefit of tension–tension bracing, as its name implies, is that the members only need be designed as ties, with only one tie acting at any one time, depending on the wind direction.

9.7 Stress–strain and the choice of material

Visualising the sets of internal forces in structural members can be considered without any reference to the size, shape or materials of the structural members. However, it is an understanding of the type and magnitude of these distorted shapes and internal forces in the members and joints that gives clues to the design of shape and materials that may be possible within economic sizes and manufacturing process.

To proceed with the analysis it is necessary to propose the shape and size of the structural member. The internal forces can now be associated with the cross-sectional area of the member over which they will act. From this it is possible to determine the stress that will be generated in the material. The stress is determined by the internal force acting over the proposed area given by the size and shape of the cross section of the member. So far, the illustrations have shown solid cross sections. For beams this has been rectangular, but the distribution of stress across the section suggests an 'I' or box section could be used. This removes material from areas where the bending forces reduce towards the neutral axis, using less material and reducing weight, but it changes the stress distribution across the section. Whatever the cross-sectional shape the stress distribution is independent of the material chosen; it is only a function of size and shape.

The choice of size and shape, however, cannot be established without knowledge of the material's structural properties, mainly (although not exclusively) associated with the stress–strain relationship. It is now that it is possible to ask questions about the two key failures associated with structural performance:

- Will it break? This is known as the collapse criterion and is directly related to the strength characteristics of the material. For a safe solution, the allowable stress in the material must be greater than the stress that will be generated by the internal forces over the area of material proposed.
- How much will it distort? This is known as the serviceability criterion and is related to the strain characteristics of the material. The distortion must be limited to maintain visual lines and ensure finishes do not crack or become detached. The structure must also not feel live (bouncy) when subject to the imposed loading.

The simple question of whether it will break for the collapse criterion to be satisfied is not enough. In the interests of limiting the risk of collapse while designing an economic structure it would be more accurate to ask, 'How close is it to breaking under working conditions?' The design must incorporate a factor of safety. This will be greater the more sensitive the design is to variability of the loading and/or the materials' properties. How this is allowed for depends on the approach taken to the design.

For every material there is a relationship between the stress experienced and how much it strains, known as the e value. This relationship is shown in the stress–strain curves depicted in Figure 9.13 where the curves for concrete and steel are shown. The stress–strain relationship identifies the stiffness of the material, how much it strains with an increase in stress, which makes some materials difficult to use in building. It may not be possible to take some materials to their full working stress without the strain induced by that stress leading to excessive distortions. Generally, building structures are made from relatively stiff materials such as masonry, timber, concrete and steel.

There is another important question in considering the collapse criterion. Not 'When will it break?' but 'How will it break?' This is another characteristic of the stress–strain relationship of the material that can be identified from the curves in Figure 9.13. It is the distinction between brittle and plastic failure. The curve for concrete shows that failure is sudden: it fails without any visual warning; it is a brittle material. The curve for steel shows that beyond a certain stress, known as the yield point, the rate of distortion is much greater as the stress increases, so visual warning of failure is given; it is a plastic failure. A second characteristic is that releasing the load after the yield point has been passed will not return the material to its original shape; it is deforming plastically as opposed to elastically.

Once the stress–strain relationship of the material is known, it is possible to evaluate the suggested shape, size and material specification for each member, ensuring that the stresses induced by the forces on the member are not excessive and the strains induced by the stress do not cause unacceptable distortions. This process of checking suggested member size and shape against induced stresses and materials properties is the basis for structural analysis and design.

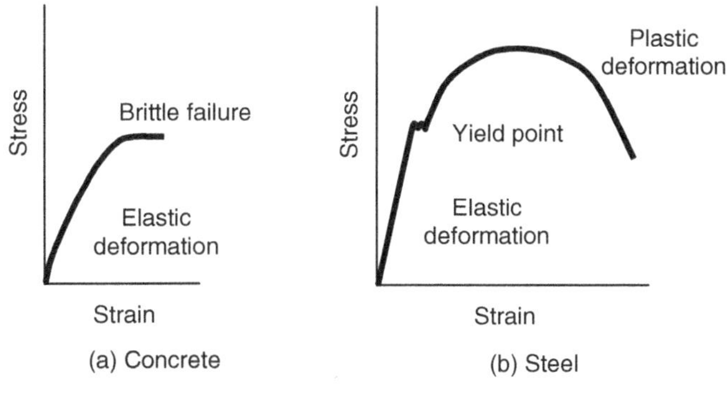

Figure 9.13 Stress–strain curves.

9.8 Structural analysis and design

Having established the loading pattern and devised a structural system of members and joints that is inherently stable, it is possible to identify the structural behaviour of each joint and member and determine the pattern of forces that will arise in each. Knowing the expected behaviour of the structure and materials under stress, it is possible to suggest appropriate cross-sectional shapes, sizes, materials and details for the structure.

It is necessary to evaluate the performance of these suggestions to see whether they can offer sufficient resistance without breaking and within limits of distortion. Initial proposals can normally be made from observations of past construction combined with an understanding of structural theory and design methods. Theories of mechanics and strength of materials support engineering knowledge. These theories are used to devise mathematical relationships that predict forces, stresses and materials behaviour sufficiently accurately to be relied upon in design. They must, however, be based on an initial ability to visualise the behaviour of the structure and the materials involved. If the engineer does not imagine behaviour and then suspects that behaviour as being a possible failure mode, the calculations necessary to check the design may not be undertaken. Many structural failures can be attributed to this error of judgement.

Once behaviour of the stable arrangement of members is visualised and the most likely failure modes identified, calculating the magnitude and distribution of the internal forces induced in the members is known as the process of structural analysis. Choosing and evaluating the shape, dimensions and materials for the members that will safely resist these forces to limit distortion, local instabilities and ultimate loss of load-carrying capacity is the process of structural design. Design also includes the detailing of the members and joints to ensure the behaviour assumed in the analysis.

9.9 Movements and structural behaviour

So far, the analysis has assumed that the only movements or change in shape or dimension from those originally formed are the distortion of the members under load. It has also been assumed that these will be able to take place freely, without any restraint from other parts of the building other than their connections one with another. There will be other movements caused by changes in environmental conditions changing the temperature and moisture content of the materials. These will not only change the dimensions of the structural members themselves but also may affect all the other parts of the building depending on the detailing of the joints and fixings. When these movements take place, they may affect the structural system in one or more of the following ways:

- Restraining this change in dimension will change the magnitude and distribution of the internal forces in the member.
- If the member must restrain these dimensional changes from other parts of the building, this will act as additional loading and hence change the magnitude and distribution of the internal forces.

- If movements elsewhere in the building result in a change in the position of some support in the system (most likely ground movements), major redistribution of the internal forces may occur (often determined by joint rather than member behaviour).

Every effort should be made to predict these movements and to detail the joints and connections to allow free movement. If this is not possible then the effects must be evaluated as additional loads and hence levels of stress.

An additional complication is that some materials, such as concrete, continue to distort with time even though the load remains constant. This phenomenon, known as creep, means that the effects of these dimensional changes may develop at some time in the future, leading to failure sometime after the building has been occupied. One important aspect of movement in the design of a building structure is that of the relationship between any differential settlements in the foundations and the ability of a building structure to move with these settlements without distress. Each overall design of a superstructure (members, materials and joints) will have an associated flexibility that will determine how much differential settlement they can accommodate between adjacent foundation supports and this will be a key performance criterion for the foundation design.

9.10 Stresses in the ground

Analysis so far has been to establish a stable arrangement of structural members, where the task has been to suggest the shape, dimensions and materials to be used and then check that the internal resistance is sufficient to carry the loads safely without distortion or breaking of any part of the structural system. The ground on which the building is founded is also part of the structural system but cannot be considered in this way. It too will have to be checked for stability (subsidence), distortion (settlement), and breaking or collapse failure. However, unlike any of the structural elements discussed so far, the ground cannot be specified, it must be investigated. The aim of the investigation is to determine the information necessary to check out the ground's ability to carry the loads without excessive settlement or collapse failure. However, it is not possible to determine the scope of the investigation or type of information required without some idea of the pattern of forces that will have to be resisted if the soil is not to fail. In this respect it is the same process as for all structural members, but the visualisation of the behaviour needs to identify the shape of the stressed area. This is very different from the structural members considered so far since there is now no obvious edge or boundary to the soil as there is for a structural member.

Figure 9.14 shows the way the soil will be stressed and shows both distortion (settlement) and collapse behaviour under a bearing foundation. Figure 9.14a gives an idea of how the stress will build up in the soil. This shows the volume of soil that will strain, leading to an ability to predict settlement.

There is a simple classification of soils into cohesive (broadly clays) and non-cohesive (broadly sands and gravels), which is useful when thinking about the behaviour of the ground. One of these behavioural differences is how long it takes for the ultimate settlements to take place. In non-cohesive soils the settlements will take place as the loads are applied. Cohesive soils often take years to reach a final settlement and this time-dependent behaviour must be investigated for a safe design to be achieved.

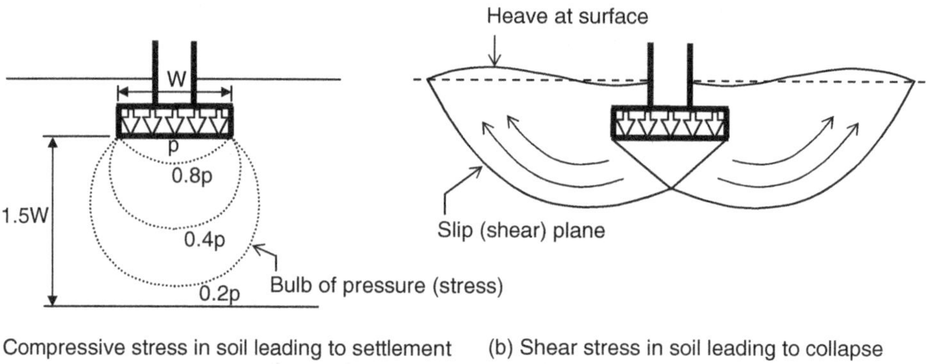

Figure 9.14 Stress in ground due to foundation loading.

The observed collapse failure in soil shows how a shear plane develops in the soil which reaches the surface some distance from the foundation. Figure 9.14b shows two symmetrical shear planes. Only one needs to develop for failure, the foundation rotating with the one-sided slip. Comparing the two illustrations, pressure bulbs and shear planes, indicates that the volume of soil involved in collapse is not the same as the soil involved in settlement. Together the two diagrams show the extent of the investigation and the type of information required to predict the behaviour of the soil. The stress–strain (load–settlement) relationship is necessary to predict settlement and the shear strength needs to be known to check for collapse. These can be identified from laboratory tests on samples taken in the investigation or from *in situ* tests taken in the undisturbed ground.

These direct limits of depth and spread of the area of investigation for settlement and collapse may miss the third failure mode, the disturbance or overall instability from subsidence. This would cause the movement of a mass of soil that removes the support for the volume of soil stressed by the loads from the foundations. Subsidence may be geological in origin, with landslips or swallow holes, or may be man-made, with cut embankments or mining activity. This also must be investigated, but the greater scope of area of investigation and the type of information required would need a desk study or topographical survey, although this may identify the need for a more widespread physical site investigation.

Another disturbance to the foundations not related to the direct stress in the soil is volume change due to changes in environmental conditions in the ground. Changes in moisture content in clay can significantly change the volume of the soil. In these 'shrinkable' clays, foundations must be taken below any risk of moisture change or the foundations may settle (or heave) independent of the stress in the soil from the building loads. Frost heave is another environmental change causing a volume change in the soil. Soils susceptible to movement under these conditions include silts, chalk, fine silty sands, and some lean clay. Other ground movements such as earthquakes generate stresses and changes in the soil properties, leading to new forces on the building (these are beyond the scope of this text).

9.11 The major structural forms

It is now possible to return to the three major structural forms introduced in Chapter 3 and see the types of members and materials that can be used economically to achieve a variety of shapes and sizes of buildings for a variety of different uses.

Loadbearing wall structures

Structures based on loadbearing walls, by definition, use walls for support. In the past this has been the main structural form that has been developed for domestic-scale construction to larger buildings carrying high loads from relatively large spans, but this generated thick walls, increasing dead loads, and limiting architectural form. The walls would have been based on masonry, probably stone or brick, or timber originally used to frame up whole walls and more recently to frame panels that can be used to create walls of varying lengths. Developments in materials such as concrete and steel have provided other options in skeletal frames and long-span roofs for these buildings but also other options for loadbearing solutions. The loadbearing solution is now perhaps most widely used where floor spans are limited to, say, 4 or 5 metres. This limits loads, giving relatively thin walls even for relatively high-rise structures. Where the building requires relatively small internal spaces on a regular pattern that repeats on all floors so walls can be continuous to the ground, loadbearing wall structures are an option. This is particularly true if the walls have sound or fire requirements, which tend to lead to construction that is good at carrying loads.

Modern, stronger and quality controlled materials have allowed buildings up to 18–20 storeys to be constructed, but only where there are many internal as well as external walls available to carry loads. Typical of these buildings would be hotels, student residences and flats. If internal layouts allow, low-rise commercial development with greater floor spans may be considered for a loadbearing solution.

Floors for commercial and high-rise buildings are likely to be concrete slabs, although, given limited internal spans in a single dwelling the timber joist (small beams at close centres) and boarding floor is probably the economic solution where sound and fire requirements are less. Roof structure is likely to be timber truss rafters (pin-jointed frame) on low-rise buildings.

If the cellular nature of the internal room layout provides walls in both directions, the resulting structure is naturally stable against wind forces. If, however, the loadbearing walls are predominantly in only one direction on plan, the building will be unstable in the other direction and some other stability members will have to be introduced. These may be from internal shafts such as stairwells or may be short returns, known as piers, at the back and front of the building on the ends of the wall to stiffen the building in the vulnerable direction. Construction where major loadbearing walls are used only in one direction is known as cross-wall or box frame construction and allows a variety of front and back façades to be chosen for the building. Single-storey industrial buildings with long-span roofs can be constructed with loadbearing walls in both masonry and concrete.

Skeletal frame structures

The development of concrete and steel as structural materials allowed the design of effective and economic bending members, particularly beams and slabs that form two of the major structural elements of the skeletal frame. Steel has high strength in both compression and tension. However, steel is expensive and to develop economic bending sections it must be formed into the classic 'I' cross section. Concrete, while having no significant structural tensile strength, has sufficient compressive strength to make a composite with steel to resist the tensile forces, forming reinforced concrete members. The success of reinforced concrete as a composite is based on more than the relative strength characteristics of steel and concrete. The strain, thermal expansion properties and cost are also complementary. The shrinkage of concrete on curing, plus some ribbing on high-yield reinforcement, provides the bond necessary for the two materials to act as a composite.

With the newly developed high-strength materials comes the ability to take vertical loads on small cross sections so walls could become columns. The basic form of the skeletal frame shown in Figure 9.15 could now be formed effectively to offer not only more clear internal space in buildings but also greater height without the greater thickness of materials that went with the limited strength of masonry. Floor slabs are the third major component of the skeletal frame. Almost universally constructed in reinforced concrete, even for structural steel frames, these floor slabs have been developed in several forms both *in situ* and precast. This gives a range of options that can accommodate spans between 3 and 12 metres, but their economy is determined by their choice as part of the frame layout.

It is not normally economic to have columns at much less than 6-metre or more than 12-metre centres. This is linked to the choice of slab. This would mean, for instance, the use of secondary beams for a 3-metre one-way spanning slab or up to a 12 × 12-metre column grid if a two-way spanning waffle slab were used. Slab developments have also led to reinforced concrete frames without beams where plate or flat slab floors have slabs transferring their loads directly to the columns. This generates high punching shear stresses around the

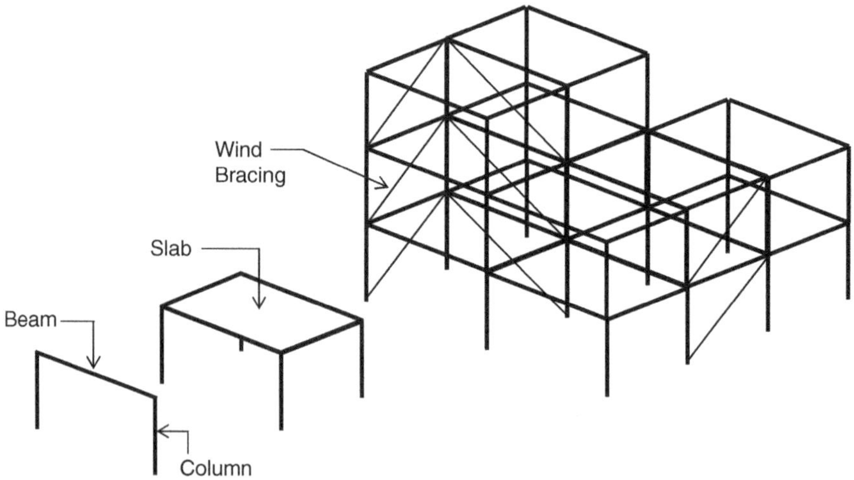

Figure 9.15 Basic members and arrangements of skeletal frames.

columns. These floors with office loading can span 9 metres at around 350 mm deep in reinforced concrete and even greater spans (or shallower depths) if pre-stressing is employed. The economy in the flat slab is achieved in the simplicity of the formwork and falsework systems, which are a major proportion of the cost of reinforced concrete structures.

Wind stability members make up the other major structural member of the skeletal frame. Shear walls are probably the natural choice for *in situ* reinforced concrete frames. Compression or tension – tension bracing may be used with steel or precast concrete frames, although concrete or masonry cores are often an alternative, depending on the layout of the buildings. The skeletal frame no longer requires external walls to carry dead and imposed loads from the building, only to transfer wind loads to the structure. This has allowed a wide variety of façade materials to be employed to give expression to a range of architectural forms.

Long-span roof structures

The need to roof over large areas with little interruption even from columns characterises many buildings. Initially developed for warehouses and factories, this form is now employed for retail and sports and leisure use. Predominantly used to cover ground-floor areas, they can also be used to form a roof over upper floors where the activities on lower floors can be accommodated in a skeletal frame structure in buildings such as airport terminals. Although long-span roofs are often thought of as having a 35-metre span, or more, the simpler plane frame, and even some of the three-dimensional forms, can be economically used over much shorter spans. Although not universally accepted as a definition, it is useful to think of these forms as being used on short spans of up to around 15 metres and as medium span from 15 to 35 metres. This gives some clues as to which forms will be economic.

While, like skeletal structures, they are based on the frame (a bending structure supported on columns), they have been developed into a much wider variety of forms. Many are expensive and gain their value in their dramatic forms over medium to long spans and are often left exposed as part of the architectural design. The simplest and generally least expensive form is the plane frame. There are now two common forms: the pin-jointed frame, such as trusses and girders supported on columns, and the portal or moment frame, where fixed joints distribute the bending moments into the columns, giving economic beam depths even over large spans. The triangular-shaped truss used to form a pitched roof is only normally used for short spans, but both girders and portals can be used for a long span of up to 100 metres if required, although smaller spans are more economic.

The spacing between the frames will be dependent on the ability to provide purlins and cladding rails to support the cladding (see Figure 9.7). On longer spans the individual frames become more expensive so setting them further apart can reduce the cost of the frames, but the purlins and cladding rails become more expensive. On short to medium spans, frame spacing of 4–8 metres with cold-rolled steel Z purlin and cladding rails can be used. On long spans, 8–12 metre spacing of frames with light truss purlin and cladding rails may become more economic.

Wind stability provision is less straightforward than for skeletal frames. Portal or moment frames are inherently stable against side wind, which will put additional bending forces in the frame (sway) that must be analysed. Being plane frames and not connected other than by purlins and cladding rails, they are unstable against end wind loading and require bracing in this direction. The long span creates large roof areas, and the light roof cladding makes these frames susceptible to wind uplift. Given that the roof sheeting remains connected to the structure, the frame may be subject to bending reversal. The top of the beam will go into tension and the bottom is subject to compressive forces. In these frames one failure may become lateral torsion buckling, as the bottom of the beam is unrestrained. In steel portals, for example, struts must be specified to the bottom flange of the 'I' beam against this condition.

Pin-jointed trusses and girders, with pin joints to the top of the column, are vulnerable to wind in both side and end loading. The end-loading wind may cause the frames to overturn not only at the connection at the bottom of the column to the foundation but also at the connection of the truss to the top of the column. Bracing for the walls and at the eaves will be necessary. As the height to eaves of these frames and/or the spacing apart becomes greater, the length of the bracing member becomes longer. If single bracing is used, the increasing slenderness makes the buckling failure more likely and large-section members must be used. If tension – tension bracing is used the increased length does not increase the chance of failure, making the two-member form of bracing economic.

The second broad type of long-span roof is the three-dimensional form. Either it takes advantage of the two-way spanning pin-jointed frame forming flat, two-layer grids such as space frames, or it gains the spanning advantage by curving (or folding). Curving can be achieved either in single-layer grids such as geodesic domes or from solid slab-like structures such as shells or folded plates. These forms are relatively stiff, but curving can be used in the more flexible tension structures based on suspended nets (cables) or fabric structures.

These three-dimensional forms can all be used over medium and long spans in a variety of construction options and materials, achieving a wide range of architectural forms, making the analysis of their choice complex, with a need to be well informed. It is not sensible to offer broad advice on form and spacing of these three-dimensional forms as it is with the skeletal frames and the plane frame solutions for these long-span roof structures. For additional information see *Barry's Advanced Construction of Buildings*.

Physical Behaviour Over Time: Chapter 10 AT A GLANCE

What? During the life of the building changes will occur in the materials and components that affect the reliability of the construction to continue to perform. The three broad changes that can bring about a loss of reliability and failure are environmental deterioration (weathering), movements and wear. The durability of the building will be influenced by the materials and components used, its orientation and level of exposure (location), and how it is used and maintained. These factors will influence the building's durability.

Why? Environmental agents causing deterioration are specific to each material, and therefore it is necessary to determine the concentration and frequency of these agents in the environments associated with the specific components of the building. This is often very localised to one vulnerable part of the detail, for example a joint between two materials. Failure to understand the interaction of the parts at a joint may lead to premature failure and the need for unscheduled replacement.

When? Because buildings are exposed to the local climate and are used on a daily basis they will deteriorate. Weathering and wear are usually relatively constant, although the signs of distress or damage to materials and components may not be evident for many months or years. Extreme weather events, such as high winds and/or heavy rainfall and surface flooding may result in damage to the less resilient buildings. Regular visual maintenance checks will usually identify surface level damage, however, issues within the building fabric will only be revealed through the use of surface penetrating radar or by opening up the fabric.

How? Movements, or changes in shape, size or position of components, can be caused by initial equalisation processes following installation as well as the loading and environmental conditions of temperature and moisture changes that may take place during the life of the building. Movements in the ground then require a separate understanding associated with subsidence, which is in addition to the loading and moisture changes that cause settlement. Wear takes place in parts designed to move and is, therefore, more significant in services design where provision must be made for the replacement of parts or whole units. Not all these changes lead to failure. Distress in the building leading to loss of performance will depend on the detailing and how it limits deterioration and accommodating movements.

10 Physical Behaviour Over Time

This chapter deals with the third major aspect of the physical behaviour of the construction: consideration of the analysis of deterioration, movement and wear. All these processes will happen during the life of the building and may reduce the reliability of the construction to continue its contribution to the performance of the building. Concerned with the changes that may take place with time, it considers the potential failures and the implications for maintenance and repair during the life of the building. Understanding the way in which a building will be subject to weathering, how it may move with the change of seasons and the passing of time, and how it will wear because of daily use will inform our analysis and choice. This requires an ability to visualise the building over its service life and understand when components are likely to need to be replaced as well as the long-term potential for material recovery and reuse at the end of the building's service life.

10.1 Reliability: renewal, maintenance, repair and reuse

The analysis of behaviour over time is about understanding possible changes that take place in the materials, components, joints and fixings that will affect the ability of the construction to perform. Changes may reduce the chance of performance the next time design conditions are experienced. The reliability of the construction is reducing. Its chance of failure under the next set of testing conditions becomes greater. However, in most cases, construction is not just built and then demolished at the end of the life of the building. Its reliability may be dependent on interventions during its life to keep satisfactory performance. To evaluate a proposal against its reliability over time it is necessary to specify:

- ❑ How long it is intended to last before it is renewed?
- ❑ What maintenance it can expect during its life?
- ❑ What is the impact of failure and the ease of repair?

To complete the specification it is then necessary to identify:

- ❑ How it can be dismantled?
- ❑ What options there are for safe recovery and reuse of materials and components?

Barry's Construction Technology: Analysis and Choice, Third Edition. Stephen Emmitt and Tony Bryan.

Without both the initial detailing and specification and a clear statement on these interventions in the life cycle of the construction, the proposal has not been fully evaluated.

Not all parts of the building must last for the same length of time before renewal. Clients' and users' requirements change and, while not foreseeable in detail, they are likely to involve finishes and the division of space. It may be more economic to renew than to maintain, although this may not be desirable against sustainability criteria. It is necessary to establish as part of the analysis the client's commitment to renewal cycles and maintenance plans. Another strategy for some parts of the building is to expect failure and renew or repair when this happens. For the client this has a cost burden for loss of use and in establishing an organisation of repair in the case of failure during the operation of the building.

Decisions on repair and maintenance concern all parts of the building, from tap washers, decorations and roof finishes to the life cycle of the whole building. Seen in a life cycle context this will include future adaption, retrofitting and change of use. Clients and building owners will have a view on the management of the building they use for their everyday operations and the amount of money and time they are willing to dedicate to maintenance and repair. Decisions on value for money and sustainability are not just concerned with the initial building costs, but the whole life cycle of the building. The basis of choice is therefore to understand the changes that take place over time, and then to consider the options for renewal, maintenance and repair to sustain operational performance.

Project briefs will have an initial specification that should foresee anticipated maintenance, with an analysis of the consequences of failure (safety, cost and disruption) and the ease of repair and replacement. This will include requirements for safe access that must be incorporated into the design. If the client does not favour maintenance, and repairs cannot be made easily, then initial specification and detailing may have to be changed before the final choice is made. Finally, there is the consideration of dismantling and recycling, which may also include decommissioning. Whether considering renewing a part, or the demolition of the whole building, the ease with which it can be taken apart and the impact of disposal should be anticipated in the original specification and details. When seen as an economic issue it may be so far into the future that the temptation is to ignore it. However, with the emphasis on sustainability and the quality of the environment for future generations, dismantling and disposal can no longer be dismissed at the design and detailing stages. It must form part of the analysis and choice cycle.

10.2 Basis of analysis

In the past the limited number of construction materials coupled with relatively low performance expectations allowed (local) knowledge to build up over the years. Details were developed to protect parts from weathering and decay. Where protection was not possible, there was a greater acceptance of maintenance activities to keep the construction in good condition. The details that developed were often regional to fully consider the conditions for the locality. The orientation and grouping of buildings considered local exposure conditions to protect the vulnerable parts of the building from the worst ravages of the weather. This resulted in a local building vernacular, with craftspeople applying their knowledge and experience to good effect. This no longer holds true for most buildings erected today. This places greater emphasis on the designer's ability to undertake an analysis of behaviour of the construction with the passage of time for a specific context. This can be enhanced by close observation of the local vernacular and analysis of how buildings around the development site are reacting to the local climate.

There are three broad areas of analysis that must be considered as significant to the reliability of building, each with a potential to lead to failure, invoking the need for maintenance, replacement or repair:

(1) Environmental deterioration (durability)
(2) Movement
(3) Wear

There is a need to understand their direct effects to undertake an analysis of the proposed solution to identify ways in which the building may become less reliable, require effort to maintain and ultimately fail.

10.3 Soiling and cleaning

Before considering in detail the three broad areas of analysis identified above, all of which may lead to physical failure, the less threatening but disfiguring accumulation of dirt on both internal and external surfaces needs to be discussed. This soiling affects appearance and may be sufficient to justify cleaning during the life of the building. Externally soiling is part of what is sometimes known as weathering, the other part being the environmental deterioration, as a visual change to the façade. Like the more destructive aspects associated with durability of the materials, soiling is dependent on the environmental conditions. These determine both the type of dirt deposited and the natural cleaning action of rain.

On a façade both deposition and natural cleaning can be quite localised, and this can lead to characteristic patterns of staining. In splash zones, where water is splashed onto vertical surfaces from roofs, ledges or the ground, soiling can be heavy. Where water runs from ledges, such as the ends of windowsills, deposition may be greater than on the general face of the façade. At the same time sheltered areas under ledges may not be washed by the rain. Whole walls either less exposed or sheltered by other buildings may accumulate dirt more than open walls even on the same building.

Some materials are more susceptible to soiling than others. This may be due to colour, where light shades are more vulnerable, or the surface provides for greater adhesion, making cleaning processes less effective. Some of the more traditional smooth renders have these characteristics. Detailing is also significant in that it can encourage both deposition and natural cleaning processes. Joints and overhanging details need particular attention. If excessive soiling occurs on a building, cleaning may have to be considered.

Windows, and possibly roof lights, have surfaces where cleaning always must be considered. While it is possible to specify a self-cleaning glass, most buildings will rely on a cleaning regime undertaken at regular intervals during the life of the building. Modern systems are now available using brushes on long poles that deliver water that is processed so that it does not need to be squeegeed or leathered off. This gives the opportunity to clean windows from the ground in a safe and efficient manner. For taller buildings, some form of anchorage for window cleaning access may need to be designed to connect directly to the structure, often on the roof. Access equipment can then be suspended to travel down the face of the building for safe cleaning, maintenance and repair.

10.4 Durability of materials

Over time, changes take place in the properties and integrity of the materials of which components are made. This will alter their performance in the future. It should be acknowledged that some of these changes are beneficial in that they improve the durability of the materials. The weathered surface of stonework is a good example of this. This makes the material more durable. It is important to recognise this not only when specifying new building work but also when considering cleaning, maintenance and alterations. Changes may reduce the durability of the material. Durability of a material is dependent on the conditions to which it is subjected. Each material has specific agents of degradation or preservation. It is not possible to determine the durability of a material in a building without a clear definition of the conditions under which it will have to operate. Each material has associated agents and mechanisms that are known to degrade or preserve the materials. Either material must be chosen to survive in the expected conditions or, unless a new material is specified, some protection or preservation must be specified.

Agents and mechanisms

The mechanisms of change can be identified as taking place either within the material or at the surface. In both cases there will be a change to the internal structure of the material that is affected. This internal change can either affect the properties of the material or disrupt the internal integrity of the material, both of which result in a change in performance, most often for the worst. Changes in properties include embrittlement in plastics and fatigue in metals, while examples of internal disruption are wet rot in timber and sulphate attack in ordinary Portland cement-based materials. In each of these examples changes only occur in the presence of specific agents. Embrittlement is associated with ultra-violet light, fatigue with reversing stresses, wet rot with dampness and sulphate attack with sulphates and moisture.

These examples also serve to bring out another useful way of thinking about these agents of change. They are chemical (sulphate attack), biological (wet rot) or physical (fatigue). There is often a coincidence of conditions as in frost disruption where saturation and cold temperatures cause a change of state, not in the materials but in the saturation water, causing a physical internal disruption of the material normally just below the surface. Water plays a part in many of the deterioration mechanisms, as does air with its ready supply of oxygen. These components are significant not only in many biological processes but also in chemical processes, for example oxidation. Many of the materials from which we make our buildings are taken from nature and they are part of the natural decay–regeneration cycle. Understanding the mechanisms and agents in this cycle becomes the key to predicting the durability of many materials and diagnosing failure.

Many of the more recently adopted materials we use are not naturally occurring but have been refined or processed. The pathways and agents of change in these cases are often to revert to their original or lower-energy states. Many processed materials have proved to be stable with time, for example glass. However, there is a tendency in some processed materials to change to lower-energy states, which may not have the properties that we carefully designed into the processed materials. Many of these mechanisms also need agents, again often water and oxygen, with other chemical agents accelerating the process. The mechanism of oxidation in metals is a good example of this.

Oxidation is a surface process where the oxidised layer does not have the properties of the parent material. Oxidation of iron (rust) is recognised as reducing the life of a component. Not only does the oxidised material have a greater volume than the parent material, which increases the dimensions of the original component, but also the rust layer ruptures and allows the agents of oxidation to reach down to new parent material, continuing the cycle of degradation until all the material has rusted away. However, this is another example where not all changes reduce durability. Non-ferrous metals, such as aluminium and copper, oxidise, but this new surface, while of greater volume, changing the dimensions of the component, is not great enough to rupture the surface layer. This new oxidised layer, or patina, is stable at normal temperatures and therefore protects the underlying parent material by significantly slowing down the rate of degradation. The component's life is potentially increased.

The identification of agents associated with the mechanism of deterioration or preservation for each material is a key aspect of the analysis of the durability of a material. It involves the identification of the environment in which the material is to function.

Cycles and concentrations

Because the rate at which changes occur in materials varies with exposure to the agents of change, it is necessary to project an expected pattern of exposure. The pattern must identify the frequency with which changes in concentrations of agents, which are in contact with the materials, are expected to occur. Because many deterioration mechanisms have several agents involved, a pattern for each must be established so the frequency of coincidence can be identified. In frost damage it is the coincidence of saturation and low temperatures as well as the frequency of the freeze–thaw cycle that is crucial to the rate of deterioration. If the material remains dry at low temperatures, there is no deterioration. If the material is saturated but remains frozen for long periods, there is little chance of deterioration. If the material is saturated and freezes at night and thaws during the day, there is a greater chance of deterioration.

It is very important to carry out this analysis of conditions locally, in immediate contact with the material. It is very possible that even different parts of the same component are subject to very different patterns of exposure. Failure is often localised, as with the wooden fence post rotting around ground level shown as a high-risk area in Figure 10.1. Here the patterns of concentrations of moisture, air and decay organisms are at their greatest and the cycles of change at their most frequent. This is also a good example to identify the preserving nature of some agents if their concentrations remain high but steady. The example here is water and timber. If water is constantly present, maintaining the timber in a saturated condition, it acts to preserve. The fence post buried deep in permanently wet ground does not decay. To continue the fence post example, the top of the post is also vulnerable as it is end grain that allows the water deeper into the timber, providing a pathway to high local concentrations of the decay agents.

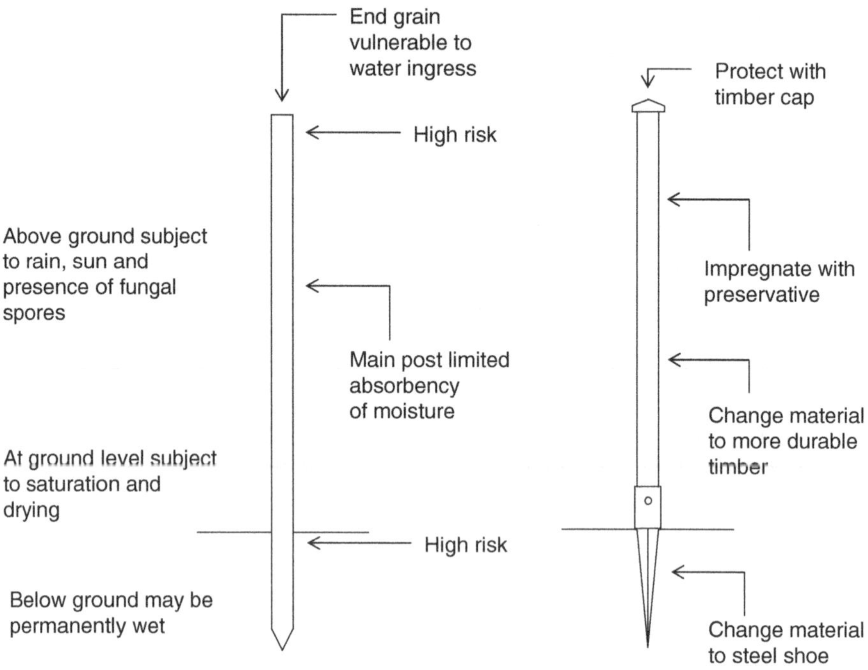

Figure 10.1 Timber fence post – deterioration and improved durability specification.

This pattern of conditions allows an analysis of the rate of deterioration that takes place at the critical point in a component. Most materials in the building will have periods of decay and periods when little change is occurring. This may be considered in the long run as a gradual change. The changes taking place are making the component less reliable the next time the component must operate at or near its design limits. The fence post breaks in a high wind but no worse than winds it has survived in the past.

Where deterioration patterns are identified as creating a high probability of failure (or the consequences of failure are catastrophic), consideration must be given to changes in the suggested specification. It may be possible to protect the vulnerable parts from the agents of decay or provide some preserving process that limits the effects of the decay agents. It may be economic to change the specification of the material. In the case of the fence post the end grain at the top can be protected with a cap, the timber preserved particularly where it enters the ground, or a metal shoe placed in the ground to receive the timber above the aggressive line of the ground. In this case the deterioration of the metal shoe would have to be analysed.

The maintenance expectations for this specification of the fence post may include the gradual reduction of the effect of the preservative, which would have to be replenished, applied by brush sometime in the future. Maintaining and/or repairing of the cap at the top of the post would be cost-effective as it is easy to do. The ultimate failure may well now still be at the base of the post even with the metal shoe. If maintenance of the metal shoe is not contemplated, it may be that repair of the first few posts to fail could prolong the life of the fence, but ultimately the renewal of the whole fence would be the preferred option. Dismantling would not be problematic, and neither would be the disposal of the timber if the preservative was not still active. It may be that at this stage the metal shoes may be cleaned, renovated and reused.

10.5 Movements in components

The term 'movement' here refers to changes not only in position but also, more generally, in size or shape of the components. These are changes in the dimensions of components. All buildings and their components will be subject to these movements, but, if correctly detailed, this need not lead to failure. Failure would be manifest in cracking or distortion. When small, these cracks or distortions may disturb finishes and allow water penetration but in the extreme they can threaten structural stability. Failure is, however, not inevitable. Movements of this kind can be accommodated and allowed for in the choice of materials and in the detailing.

The dimensional changes are often termed inherent deviations as they occur due to the inherent properties of the materials. This contrasts with the induced deviations that are the result of the production process and need to be considered to ensure that the initial construction will fit together. Many dimensional changes pose little threat to the building. If they can take place freely, without impacting on adjacent construction, they will have little effect. Movements that are of concern are generally differential, where one part is moving more than another to which it is connected, or restrained, where one part is moving in one direction and meets another that cannot move (or is indeed moving to meet the first part). However, some movements need be very small to have a large effect. The shrinkage of cement mortars in brickwork breaks the bond with the brick, causing fine capillary paths leading to the potential for damp in the building, where the most common solution is to introduce a cavity to break the damp path to the inside of the building.

Much of this movement can be accommodated in the normal detailing if care is taken in thinking through the probability that induced deviations will not eliminate the gaps and/or that the limit of flexibility of jointing materials that allow the movements to take place freely is not exceeded. These details then must be built with care to ensure that the expected allowance is achieved. Some movements are so significant to the integrity of the construction that specific movement joints must be designed. These may be required where structures of different heights come together with different foundation settlements. Another example is the design of large panels of brickwork where environmental changes will cause internal stress in continuous construction. The origins of these movements can be traced to one or more of the following:

- Initial equalisation immediately after installation
- The application of loads
- Changes in environmental conditions
- Volume change associated with deterioration
- Ground subsidence

Initial equalisation movements

Many materials undergo dimensional changes early in their life while the material is developing its final properties, often through changes in temperature or exchanging moisture. These movements are normally irreversible. In processing and manufacture, many materials must be prepared to make them workable to produce the shapes and forms required in the component. For some components made off site, at least some of these equalisation changes in dimensions take place before they are delivered. Many components, however, have not completed this change when built into the building, and for work carried out *in situ* all the dimensional change takes place in the final position.

Materials most often built into the building while this change in dimension is still taking place include timber, concrete and clay products such as bricks. Timber components are manufactured at moisture contents above those that will eventually be achieved in the occupied building and therefore will continue to shrink after installation. The equalisation movements in concretes, plasters and brick also include changes in moisture but in different directions. Concretes and plasters need water for setting and workability and then dry out and, therefore, like timber, shrink. Bricks that are fired and therefore very dry at the end of the manufacturing process take in moisture and therefore expand after the initial cooling shrinkage has taken place.

The important factor is often the timing of the installation. Concrete blocks laid too soon after casting will continue to shrink, causing tensile forces, leading to cracking in the wall. While most of the expansion in bricks takes place in the early weeks after firing, they continue to expand well into the initial life of the building. Some continuing expansion of brickwork should be anticipated. As brickwork normally has sufficient internal compressive strength this expansion may lead to the whole panel sliding on the damp-proof course (DPC) or cracking at short returns. With *in situ* work, all the initial equalisation takes place in position. Drying shrinkage in concrete laid in large areas such as ground floors can lead to cracking due to internal tensile forces that build up during the equalisation phase.

Loading and environmental movements

These movements are dimensional changes dependent on the properties of the materials and conditions experienced by those materials during the life of the building. Movements caused by loads will depend on the stress–strain relationship of the materials and on the distribution and magnitude of the loads themselves. The two major environmental agents are temperature and moisture. Nearly all materials change dimension with change in temperature. The amount of movement that takes place in a material for 1 °C change in its temperature is its coefficient of linear expansion. Only permeable materials will be subject to moisture movement. Like temperature changes, the change must take place within the material, not just in the external conditions. The amount of moisture movement that can be expected is normally quoted as a percentage of the original size.

The timing of movements caused by environmental and loading conditions can be seen to be associated with two phases that may overlap in time. The first phase takes place predominantly during construction and in the early stages of occupation, where the movements are usually not reversible. These are taking place at the same time as many of the equalisation processes and are initially associated with loading. Structural members are gradually loaded (dead and imposed) as the building is constructed and during the initial phase of occupation. This loading increases stresses that increase strains, causing distortion. In addition, the temperature and moisture conditions can be very different during construction depending predominantly on the time of year. This determines not only how quickly equalisation processes take place but also the size of components at the time when adjacent construction is installed.

Once the building is occupied, most of the loading is completed, but the internal conditions are now being established for the first time. This will move the temperature and moisture to working conditions, and during this period more dimensional changes will be taking place. Not all strains from the applied loads take place immediately. While settlements in non-cohesive soils will take place as the loads are applied, consolidation in clay soils may take years to complete, meaning that not all foundation settlement is experienced in the initial loading period. Some materials, including concrete and timber, exhibit creep. This is a continuing strain under constant sustained load. The rate of strain decreases with time but can manifest problems some years after the construction of the building. This period could cover the first few years of occupation.

The second phase in which movements may take place covers the rest of the life of the building, after the first few years of occupation. These movements are usually reversible. Most of the equalisation process will have taken place.

The effects of environmental changes in moisture and temperature on external materials become significant in this second phase. Exposed areas can experience wide variations in temperature. In temperate climates these are not so acute daily but are experienced in the annual cycle of the seasons. While air temperature varies from summer to winter, materials exposed to the radiant effects of the clear sky experience the greatest surface variation. Mid-day summer sun can raise the temperature of surfaces, if dark in colour, to temperatures of 60 °C. Conversely, clear night skies in winter will reduce temperatures of surfaces directly exposed to the sky to −20 °C, again particularly if dark in colour. This potential 80 °C difference will significantly alter the dimensions of many of the common materials used externally on buildings. Particularly vulnerable are continuous roof coverings, leading to the specification of white-coloured protection to limit the temperatures experienced by the roof covering.

External materials also experience wide changes in moisture, from prolonged periods of sunshine to extended periods of rain. Perhaps this is most often noticed in timber doors that 'stick' in wet weather. The materials affected, which must be permeable to absorb the moisture, fall into two categories: rigid internal structures and elastic internal structures. These are broadly ceramics such as brick and concrete and organic materials, predominantly timber, respectively. Ceramics are subject to relatively small changes, while timber is subject to significantly larger changes. Timber shrinks and expands differently in a different direction from the grain, changes being greatest across the grain.

Interior components are subject to less variation in temperature and moisture unless the activity undertaken in the building creates variations in these environmental conditions and no specific measures such as ventilation are taken to limit the areas affected. It should not, however, be assumed that all internal components will remain stable. Windowsills are particularly vulnerable to sunshine through the glass, raising their temperature, and fittings associated with heating systems will be subject to wide changes in temperature. Pipes should be able to move freely where they pass through the construction, or they may 'creak' as the movement takes place.

Movements due to temperature and moisture changes only take place if the changes occur within the material, not just at the surface. However, surface changes can create internal stress as the surface tries to change but is restrained by the mass of the material. This image of layers of construction expanding and contracting at different rates is particularly important if layers of different materials are to be bonded together to act as a composite. If thermal expansion properties are significantly different, changes in temperature from those at the time of bonding will create bending stresses in the materials and shear stresses along the bonded surfaces. This is also true for moisture changes. Materials bonded together that get wet can suffer similar stresses to those suggested by temperature changes if their moisture movement characteristics are significantly different.

Volume change associated with deterioration

In considering durability earlier in this chapter, some of the changes were in the properties of the materials, but some involved changes in volume. They were mainly associated with the chemical changes, such as sulphate attack or corrosion, but would also include frost disruption. These changes in volume create the potential to disrupt the component, but the overall dimensional changes have the potential to create additional stress in adjacent components with which they are in contact. An example of this is the attaching of stonework using iron fixings that rust and eventually fail.

10.6 Ground movement, settlement and subsidence

The ground is subject to two types of movement: settlement and subsidence. Settlement is the result of loading and moisture changes in the soil and therefore its analysis is the same as that for loading and environmental changes of any other part of the building. In assessing the consequences of settlement it is important to remember that it is the overall flexibility of a building that determines the amount of settlement it can sustain without distress. Brick structures in cement mortar can take less differential settlement without cracking than brick in lime mortar. Frames with pin joints can take more differential settlement than continuous structures without a redistribution of stress that may cause failure.

To assess settlement, soils are often characterised as cohesive and non-cohesive, representing the clays and the granular soils respectively. This is a useful distinction when considering movements in the soil carrying the building. Under loading conditions, clay takes considerable time, sometimes years, to compress (the action is consolidation) and even then, may be subject to creep. Granular soils compress (predominant action compaction) almost immediately the load is applied and do not change unless the loading is changed.

Clay soils undergo high volume changes with changes in moisture, not a characteristic of granular soils. Processes of desiccation in clays include seasonal drying of soil in the summer and removal of moisture by vegetation, particularly trees. These movements can be very large and very powerful. Not only will a drying and therefore shrinking clay cause additional settlement but also clay taking up moisture will heave with sufficient power to lift the building. Generally, foundations should be established sufficiently deep to be below any chance of these movements affecting the support of the building. Piles may need to be sleeved, as heaving clay will grip the pile and lift the whole pile with the building on top of it.

The freezing of water within the structure of the soil can also cause heave in some soils, typically silts, chalk, fine silty sands, and some lean clay. Foundations should, therefore, be placed below the frost line in these soils (unless building on permafrost, a condition found in some regions of the world).

Increasingly buildings are being built on ground that is neither cohesive nor granular as defined by geologically laid down material. Landfill, be it from excavation, demolition or domestic waste, is creating new development sites. This material is highly variable and cannot be easily generalised. It may have been well laid with some measure of compaction or it may have been just tipped. The engineering properties of such materials must be carefully investigated before foundation decisions are taken. Taking deep foundations through the fill, 'floating' the building on a raft or improving the ground to allow shallow foundation in the fill may represent the economic answers. This will depend not only on ground conditions but also on the building structure, considering the scale of development and the flexibility of the construction.

Subsidence causing a loss of support to the soil acting as the foundations can occur with geological changes or from human activity. Geological changes can be deep, such as earthquakes, which, while a real threat in some parts of the world, are not a design consideration in many others. Geological processes can also be surface actions, with slope stability being a particular threat to buildings. Natural slopes can become unstable in extreme weather conditions, but man-made slopes created as part of the development process can also become unstable, threatening the support to foundations. If slopes must be steep, retaining walls may have to be built to protect the buildings at both the top and bottom of the slope.

Mining activity gives a particular subsidence problem. When some mine workings are abandoned, they may collapse, causing a lowering of the ground surface in a wave behind the collapsing galleries of the mine. Buildings that exist as the wave passes are vulnerable, but after it has passed new development is less threatened. Shafts and swallow holes (natural shafts which open in limestone areas) can also pose a threat, so site investigation and a search of the geological and mining records are essential in areas known to be active.

10.7 Movement and detailing

Movements occur because of changes in shape, size or position of the components that may or may not in themselves create any damage. The possibility for damage often lies in the detailing, especially the way the parts are connected and fixed. As soon as parts are connected or touching then there is the potential for the movement or its effects to be transferred. If the detailing allows the movement to take place freely, only the component need be considered. If no such freedom can be identified in the detailing, the movement will be restrained, and force will develop between the moving and restraining components. This force may move adjacent components or may induce stresses that distort parts that are still held in position; finally it may induce stresses sufficient to crack or break components or the fixing between components.

Estimating possible movements in a particular component is, therefore, only the start of any analysis. It becomes necessary to be clear on the details to identify free and restrained movements. The first part of the analysis of details involves estimating the possible position of the parts when incorporated into the building. This cannot be assumed to be the position shown on the drawing. Each operation on the site can only be achieved within certain dimensional limits associated with induced deviation. The exact position within these limits cannot be known at the time of specification. It can only be assumed that construction will be achieved somewhere within the extremes of these allowable tolerances.

Movements take place from this initial installed position. Equalisation changes that have not taken place at the time of installation will cause movement, loading as it is applied will cause movement, environmental changes may induce movements and later deterioration may create a change in dimensions, again causing movement. Not all of these will always be cumulative. Expansions may tend to cancel out shrinkage, as with concrete where operational thermal expansions are greater than equalisation drying shrinkages. The shrinkage in concrete may, however, aggravate other details such as in the connection of a clay brick façade, which will tend to expand while the concrete frame will almost certainly shorten by both concrete curing and subsequent loading (including creep).

If this analysis predicts failure, modification needs to be made to the specification or the detailing. Materials with different properties may have to be considered. Protection for the conditions creating the movements could be considered. However, the most appropriate option may be the redesign of joints or connections to accommodate the movement, allowing it to take place in a controlled way where the detailing can include gaps or materials capable of taking the movements without loss of performance.

10.8 Wear of components

The section on movement has concentrated on the changes in dimension, shape and position that occur in components when loading and environmental conditions change. Wear is also the result of movement, but in this case in parts that are designed to move, such as hinges or moveable partition tracks. Wear is not determined by the conditions to which the components are exposed but by the frequency of operation. It is only a small part of the analysis of most construction because most building construction is concerned with passive performance. However, it is important to anticipate the amount of wear internal elements are subjected to by the building users. This will help to establish maintenance plans.

Wear is more of a consideration in services where moving parts are often associated with motors and machines that are part of services systems. Here the possibility of loss of performance at some future date requires safe access to remove and replace parts or even replace the whole unit. This must be considered in the design of the building as well as the design of the parts that make up the whole. The regular maintenance and inspection of services will identify preventative maintenance requirements, which should go some way to mitigate unexpected failure of services.

Interface Design: Chapter 11 AT A GLANCE

What? Interfaces are omni-present in a building. They range from the major interfaces between elements, such as windows and walls, through to smaller ones, for example between a fixing and the material it is connecting by way of a joint. The interfaces between the major elements of the building are the structure, the enclosure and the services. Before considering the major technologies available for the elements it is necessary to have a view on how they will interact with other parts of the building. This is a significant aspect of deciding on the technological mix at the early stages of design to have the confidence that the detailing can be resolved for the building to perform as intended. This may also have a bearing on the extent of offsite manufacturing and prefabrication.

Why? The enclosure–services interface needs to be considered very early in the conceptual design process. This interface is concerned with gains and losses flowing across the enclosure that needs to have balanced inputs from the services to maintain a comfortable and healthy internal environment. While the combination of passive fabric and active services has always needed to be balanced to maintain the internal environment, passive design has made the analysis of interaction more significant. So too has the need to reduce the carbon footprint of our built environment. It is the interface design, the joints and fixings, that will determine the ease of construction, ease of replacement and ease of disassembly.

When? The enclosure–structure interface is one of physical connection that must be resolved at the detailed design stage. Not only is the enclosure–structure interface a focus for performance with load transfer, environmental behaviour and the accommodation of movement but also is a major connection at which production issues are resolved in terms of resources, health and safety and tolerances. The structure–services interface revolves mainly around space, particularly in floor zones but also when considering floor slabs and the position of columns that can affect vertical distribution, fire safety and thermal mass.

How? Understanding the main principles of construction as outlined in this book will inform the process of analysis and choice. Relying on previously used construction details and techniques may help to limit the risk of failure, however, this is not always possible or desirable. Selecting details from BIM libraries also needs to be done while conscious of the potential interactions that may occur at the interfaces and joints.

11 Interface Design

This chapter considers the nature of each of the interfaces between the major elements of the building, namely the structure, the enclosure and the services. Before considering the major technologies available for the elements it is necessary to have a view on how they will interact with other parts of the building. This is a significant aspect of deciding on the technological mix at the early stages of design to have the confidence that the detailing can be resolved for the building to work as a whole. It is the interface design, the joints and fixings, that will determine the ease of construction, ease of replacement and ease of disassembly. This task can be facilitated by BIMs that detect clashes and interface issues. However, there is the need to understand the technologies and the potential interactions rather than rely entirely on the software to deal with interfaces.

11.1 Interfaces

Chapter 4 describes the main factors relating to materials, shape, size, relationships and the joints and fixings. It is the way in which the interfaces are designed that has a significant impact on building performance. It is, therefore, necessary to choose technical solutions for elements of the building with defined performance requirements, and these have been broadly characterised in this text as enclosure, structure and services. For most buildings these will be selected from common forms that offer the best opportunities to achieve performance and then need to be detailed for the specific building. One aspect of this selection process is to ensure that the emerging mix of technologies being considered for the building are compatible and will work with each other to ensure the performance of the building. This compatibility is not only in performance but can also be in achieving detailing and specifications at the detailed design stage. It is essential to focus on the design of the interfaces as well as the elements to ensure the building performs as intended. The design issues around these interfaces vary and must be considered at different stages in the overall design of the building.

The interface between enclosure and services is predominantly conceptual and needs to be considered at the very early stages in the design. This is when the strategies for passive environmental design must be embedded as the shape, orientation and configuration of the building is emerging. Taking advantage of these fundamental elements of design to provide passive environmental design requires early identification of the possible technical solutions for the enclosure (and structure), as these will have an impact on the type

Barry's Construction Technology: Analysis and Choice, Third Edition. Stephen Emmitt and Tony Bryan.

and extent of services that will be needed to ensure the provision of the required internal environment. There may be some physical and detailing interactions between services and enclosure where services pass through or are fixed to enclosure elements, and these will have to be resolved at the detailed design stage.

The interface between enclosure and structure is predominantly one of physical connection, which must be resolved at the detailed design stage. The most complex of these connections is between the external walls and a framed structure. Some care is needed in resolving connections to the horizontal enclosing elements such as floors and roofs.

The structure–services interface is often about space and the need for vertical and horizontal distribution in shafts and voids but, as with enclosure, services must pass through and be fixed to the structure where some services components are large and may include additional loading. However, the consideration of the interaction of structure–services design is increasingly involved with passive design, where the thermal mass of the structure can be used to store and release heat at appropriate times in the diurnal–nocturnal (24-hour) cycle.

11.2 The enclosure–services interface

There has always been a need to consider the contribution of the enclosure and the environmental services to creating and maintaining the internal environment of a building. Concerns for the global environment have caused a reassessment of this balance between passive and active technologies. It is no longer acceptable to just make the enclosure watertight and then solve internal comfort conditions with active, energy-thirsty, services. The move to net zero carbon and low environmental impact buildings has a considerable influence on the development of technical solutions for both external enclosure (roofs and walls) and environmental services. The idea is to use the passive fabric to its maximum, as, by definition, it needs no energy to moderate the environment during operational conditions. This involves using the passive fabric of the building to moderate the flows and transfers between inside and outside to maintain a quality of the internal environment throughout the year. While this approach seeks to limit the use of energy during the operation of the building, the construction to achieve this is not without environmental impact, and this must be considered in the analysis. All this has an influence not only on the choice and detailing of the fabric but also on the provision of environmental services within the building.

The enclosure and passive design

To understand the performance and selection of the technical solutions for the enclosure there is a need to return to consider what makes a quality environment. Particularly important are the elements of the thermal, light and air quality, as these are aspects of the services solutions that may have high-energy inputs with the potential to contribute to CO_2 emissions if the energy source involves the conversion of fossil fuels. These services solutions include systems of heating, lighting, ventilation and air conditioning. Systems vary in their dependence on energy but any reduction or elimination of the need for these systems will have a considerable impact on the need for energy and CO_2 emissions. The technical mix that may be appropriate for an individual building must be considered at the concept design stage, where the available fabric and services technologies need to be considered and matched to achieve economic, effective and environmentally acceptable solutions.

Passive designs for thermal environments vary depending on the daytime heat gains in the building. For dwellings these are small and there is little contribution to winter heating and a limited risk of overheating in the summer. In these circumstances the enclosure needs to restrict the loss of heat in winter, normally with thermal insulation and controlled ventilation, while maximising the opportunities for heat gains during sunny winter days. If the enclosure is effective in capturing radiant heat gains (solar gain) in winter, it may overheat in summer when heat gains are potentially high. This must be considered in the choice of enclosure, particularly windows and roof lights facing south (in the northern hemisphere) where heat gains are greatest. Windows are a key element in providing light, so the requirements of light, heat loss and utilising heat gains from the sun makes them an important aspect of the design. Some form of shading may need to be incorporated in the design of the window, particularly on the south elevation, such as shutters. For both summer and winter conditions it may be useful to consider storing heat to smooth out the effect of the day–night cycle where external temperatures vary. Capturing heat during periods of heat gain (normally daytime) and then releasing it during the night (into the building during winter and out from the building in summer) will require the fabric to have high thermal capacity. This requires some consideration of air movement and ventilation.

For buildings with high daytime internal heat gains (e.g. offices, retail and leisure) the same basic mechanisms are exploited but the focus is on limiting heat rise in summer to eliminate the need for air conditioning (easier to achieve where the natural climate does not include high humidity). This requires limiting heat gains through the enclosure, perhaps with thermal capacity and with shading to the south-facing windows, and then storing the daytime heat to be released to the outside during night-time, when air temperatures are lower. Ventilation is now important. During the day this will limit heat rise and during the night it is required to draw out the heat from where it is stored in the fabric. However, these two ventilation patterns may not be the same.

Typically the daytime ventilation needs to be around body and head height while at night the air flow needs to be across the fabric storing the heat, normally the soffit of the structural slab at ceiling level. In addition it may be possible to allow the occupants to control the ventilation in winter (opening windows), but the summer night purging will have to be automatic with windows opened automatically on a time and/or temperature basis. Ventilation can be provided mechanically as one of the active building services. However, this requires energy, and therefore a passive ventilation system may be preferable.

During the winter, daytime heat gains may not be sufficient to heat the building, although this will be dependent on the levels of insulation provided by the fabric. Heating can be supplemented with heat gains from the sun. This will be achieved through the windows, so the summer shading, designed to inhibit heat gain, must be ineffective during the winter. This can be achieved by taking the variation of the elevation of the sun above the horizon between summer and winter. Lower in winter, the sun's rays need to be able to enter the building below or between the shading provided for the summer, when the sun is higher in the sky.

This consideration of the thermal environment involves the use of the windows. These are a major part of the passive fabric involved in lighting the building. Lighting is also a major user of energy, particularly in commercial buildings. The other aspect of lighting is the depth of the building, the distance from the external wall to the back of the room to be lit. Depth must be useful for the activity in the building, and so large windows, particularly with high head heights and views of the sky, are required if the need for artificial lighting during the day is to be limited. Quality of light varies between the north and the south. North light is more even, while high levels of light in the south make glare a significant problem.

Services systems

The services systems that are required for the technological mix within a passive design must also achieve the design criterion for CO_2 emissions and other aspects of low environmental impact. Efficiency in the services equipment can help, as can automatic sensors and building management systems. Automatic control of heating systems with thermostats, and light sensors for lighting are examples of sensors, which can be more effective with zoning. Heating zones on north- and south-facing parts of the building and lighting in banks back from the windows are examples of effective zoning to match the need for the system to become operational based on environmental and user needs.

11.3 The enclosure–structure interface

The nature of this interface is significantly different, dependent on the basic form of the structure. If the structure is loadbearing, the structural elements become part of the enclosure and must be solved together. For structural frames, the interface is a physical connection, or junction between the structure and enclosure and is concerned almost entirely with joints and fixings and the shaping of components to accommodate them. Each element is likely to be created from components in different materials that only act as one entity if care is taken in thinking through how each is fixed and how joints are formed to allow the process of assembly and movement, and to give continuity of performance.

One of the characteristics of buildings with framed structures is the separation in design and performance terms of the structure and the enclosure elements, particularly the external walls of the building. This separation allows a great range of façade treatments to be considered but all will be dependent not only on their ability to modify environments, but also on their connection to the frame. This makes the focus of the enclosure–structure interface one of connection, a combination of fixings and joints to allow safe assembly on the site and to ensure the performance of the building as one entity.

With the separation of enclosure and structure the external walls are often termed non-loadbearing, but this is perhaps too simplistic. It is true that the external walls do not take the dead and imposed loads from the building. However, they are subject to wind loading, and this must be safely transferred back to the structural frame. External walls do, of course, have the more obvious performance requirements of appearance implied in the word 'façade' often used for these walls, and the environmental performance where the term 'skin' is used to indicate the external protective and environment-modifying nature of these walls. All of these, particularly load transfer, must be considered in the connection.

The main concerns in providing this connection to the frame for most of the common forms of external wall will be structural and environmental performance. To this must be added not only the major considerations of the joints and fixings to facilitate assembly at the production stage but also behaviour over time, particularly movement. For some forms of external wall such as structural glass the appearance of this connection can become a significant aspect of the design.

Of almost universal concern in providing these connections are:

- Load transfer (support and restraint)
- Environmental performance continuity
- Accommodation of movement (inherent deviations)
- Manufacture and assembly tolerances (induced deviation)
- Assembly resource and sequence, including health and safety risks

Each of these concerns is discussed in general terms below.

Load transfer (support and restraint)

Support must be provided to ensure that the weight (dead load) of the wall is taken safely to the foundation as the first condition of stability. Restraint is then concerned with horizontal forces, mainly the wind loads. However, there is accidental loading, such as explosions.

It is possible to visualise the external wall supported on its own foundations with restraint provided by the frame at appropriate intervals all over the height of the wall. While this option is often used with any heavy wall construction to a single-storey industrial building, multi-storey walls would normally transfer their dead load to the frame at each floor level.

If the support and restraint is to be provided at each floor level, the connection of the enclosure elements to the structure must be designed to take these forces. The connection must ensure the stability of the enclosure wall and transfer the loads to the structure. The nature of this connection depends on the structural behaviour of the wall. If a masonry or stud wall is specified, the wall has the strength to act as a panel spanning between the beams and columns, although additional column supports are sometimes necessary, where they are known as wind posts, acting to transfer wind forces to the structure but not carrying the vertical loads from the building as true columns. Figure 11.1 shows a schematic arrangement for the connection of this type of infill wall to the structure. It shows the simplest of arrangements for the structural element of the wall as an infill standing on the edge beam of the building structure. It is possible to conceive of this infill wall being the façade material, but it leaves the beam exposed to view. It is possible to add support components to the front of the beam to take the external leaf of the wall to cover the beam and make the façade material continuous. It is also possible to use the infill wall to provide a backing to some form of facing material. In all these cases there is a need for restraint at the top of the wall normally from the underside of the structure, where it is usual to provide a movement joint, for the reasons discussed below. Restraint will also be required at intervals in the length of the wall from the columns and any wind posts that must be provided.

Figure 11.1 also shows the wall constructed from what are normally known as cladding panels. Normally pre-formed, the panel either hangs or stands on the beam. Spandrel panels (not full height) are restrained at the columns or short wind posts, while storey-height panels are restrained at the beams and to the adjacent panel. The third option shown in

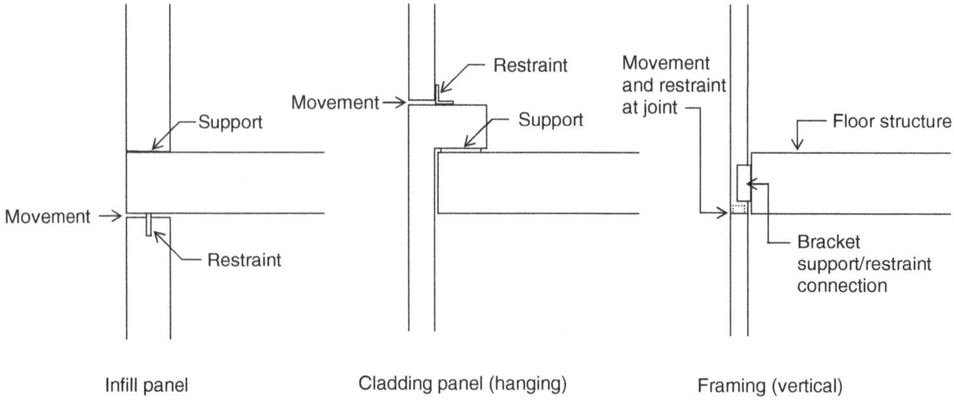

Figure 11.1 Load transfer options from external enclosure to structure.

Figure 11.1 is to create a framing to support elements of the wall that cannot in themselves span between the building's structural elements and therefore need some intermediate support. With this type of wall the interface is between the structure and the framing not requiring the continuous support of the infill and cladding panels. This point connection is most likely to take the form of a bracket capable of attaching the framing and transferring the loads to the structure.

Environmental performance continuity

The external wall of the building will have many environmental functions that may be compromised at the connection to the structure. These will include the fundamental weatherproofing requirements, particularly around damp-proofing if the basic wall construction is attached to the structure as with the normal detailing of the cavity wall. In this detail the fundamental damp-proofing is associated with the cavity and if this is interrupted at the structure, which it often is, cavity trays and weep holes must be introduced. If the damp-proofing is associated with walling systems such as rain-screening designs, attention to detailing will be associated with the connections to the backing wall, and this will become part of the detailing of the wall system more generally.

When analysing the interface connection, all environmental performances need to be considered. For a wall with high insulation values, if the structural connection brings the wall and structure physically close, there is a potential that thermal bridging will reduce the overall thermal performance of the wall. Detailing to keep the wall away from the structure to allow thermal insulation to be fitted across the structure may be an option. Another environmental performance that could be compromised by the rigidity necessary for structural connections and the space between the wall and structure is noise flanking paths. These are most acute where a physical gap is generated in systems such as structural glass walls.

Accommodation of movement (inherent deviations)

The requirements of the structural connection will inevitably require some points of fixity where relative movement cannot take place. Any change in dimensions or position in the structure will take the wall with it, and any change in dimension of the wall will be restrained, putting stresses into the wall and the structure, but perhaps most significantly the connection.

The changes or movements that take place in the structure and the wall come from different origins and can take place in different directions often exaggerating relative movements. The structural movements and distortions come primarily from the loadings. Beams bend and columns shorten. While most of this movement takes place as the loads are applied and only becomes a problem if loads are applied after connections are made, some movements are time dependent. Many foundation settlements take place over time and concrete will creep over time even with constant load. Cracking or loss of connections may not take place immediately but may be failures at some time during the operational life of the building.

The wall will experience changes in dimensions due mainly to environmental factors, although some materials will have time-dependent equalisation changes such as bricks and blocks that take some time expanding (clay) or shrinking (concrete) and may continue after they have been installed if used too soon after manufacture. There is a particular risk if a concrete frame is clad with clay brickwork. The frame shortens and shrinks while the brickwork expands. This requires positive movement joints immediately below the support of the brickwork at each floor level.

The environmental changes include both moisture and temperature changes, particularly for the cladding materials. These will depend on exposure, where perhaps the greatest risk is on the south face where the radiant heating from the sun will create surface temperatures considerably higher than the air temperature. Different materials have different properties associated with their inherent dimensional deviations. The thermal expansion of metals, particularly in thin sections where they heat up quickly, may need attention in the detailing on large areas of façade. These movements can be significant in the length of the wall as well as between the floor interface connections.

Manufacture and assembly tolerances (induced deviation)

The wall structure connection is a major point of assembly in the construction process. The structure is often produced by a different process from that of the wall components as extreme as *in situ* frame to large, prefabricated panel walls. The influence of these processes on the achievable deviations is great and the differences must be accommodated in the detailing of the connections. The wall must be attached to the frame in a way that maintains the line, level and plumb of the wall and ensures the components of the wall can fit together to maintain the performance of the wall, as discussed earlier in this book.

Assembly resource and sequence including health and safety risks

As a major point of assembly the connection between wall and structure is the focus of many of the site operations. It also takes place at the edge of the building and often at height. It can involve large, heavy and possible awkward and even fragile components. This can pose significant health and safety concerns leading to major temporary works and lifting plant. The detailing of the whole interface will determine the sequence of operations and where an individual will have to stand to make the connections, and the direction from which the components will have to be brought up to the structure to be fixed. All of these have a great influence on safe working in the most cost-effective way, both in temporary works and the time to complete the works to a safe and permanent stage of construction, minimising the need to return to complete operations later.

Poor detailing associated with this need for resources cannot be rectified by good management at the production stage. The options for sequence and resources are set by the detailing. Analysis of the assembly and disassembly process at the design stage, and modification of the detailing of this interface can make a significant contribution to safety and cost of the assembly and disassembly process. This can be done virtually in BIMs to identify any clashes and maximise production efficiencies.

11.4 The structure–services interface

The structure–services interface has always been about space and support. The most convenient place for the horizontal distribution of services systems is in the zone below the floor and above a suspended ceiling. This interacts with the floor structure, specifically with the beams. If all services run below the beams, the floor zone can become deep, particularly if ducting is involved. Flat slabs, where there are no downstands below the soffit of the slab, give a clear area for services. Where steel beams hang into the void above the ceiling, it is possible to create holes in the beams using specially fabricated cellular, or castellated, beams, although these are a deeper option than standard universal beam sections. If standard sections are used, holes can be detailed, although some care must be taken in the size and position of the hole and particularly in areas of high shear where web stiffeners may be required.

The vertical distribution of services takes space away from the usable space in the building. However, the vertical distribution also requires holes or openings in the floor structure, and these are often close to columns where the usable space is already compromised. This can produce the need for openings in areas of high shear and will give some detailing difficulties, for example with some forms of precast flooring systems. The holes also need to be detailed in such a way as to provide the necessary fire protection to prevent the spread of smoke and flames.

Other concerns around the interface between structure and services are the need to provide fixings and the more general issue of loadings. Most services systems add little to loading, but some of the mechanical plant components can be large and heavy, although from the point of view of the services design, they are usually best placed on the roof. This can cause challenges with safe access for future maintenance and repair.

The structure can be effectively used for thermal capacity to hold heat during the day for release at night. This is known as the thermal mass of the structure, which is an important aspect of passive design. This is yet another example of the interaction between the major elements that need to be considered at the early stages of design when the broad technological mix is being decided.

11.5 Detailing interfaces

As demonstrated throughout this book there is a considerable amount of knowledge required to be able to analyse and choose suitable solutions for the proposed building. This knowledge allows the building to be detailed ready for production and assembly. It is at this stage when initial design decisions and some of the detailing may be challenged to ensure that the details are suitable for construction, i.e. that they can be assembled or constructed safely with the resources available. Thus the detailing interface can also be viewed as the coming together of the conceptual and the pragmatic, the joint between design and production. This is addressed in the next chapter.

Manufacture, Assembly and Disassembly: Chapter 12 AT A GLANCE

What? Performance, quality, cost and time are determined by the sequence and methods of assembly and subsequent disassembly at an unknown date in the future. Therefore the analysis of production must be an integral part of the choice of the construction materials and details. Broad production options are characterised as being *in situ*, traditional and prefabricated. These are determined by the building design, its specification and detailing, and the needs of the client in relation to speed and cost.

Why? Analysis of the production aspects of manufacture and assembly is necessary as part of the evaluation of a suggested form of construction to ensure that the potential performance can be realised. The analysis is based on visualising the building in pieces and at the various stages of assembly and disassembly at a future date. This can be done using building information models (BIMs) to visualise the assembly and disassembly sequence.

When? The primary resources for construction work remain materials and labour supported by temporary works and appropriate levels of mechanisation. Because labour remains a key resource in production, health and safety is a major consideration in choosing methods and resources, both of which can be influenced by the choice of construction materials and details. The decision to use *in situ* or offsite production is usually taken during the briefing or early conceptual design stages of a project. These major decisions influence the range of materials and technologies that can be used.

How? The knowledge to carry out an analysis comes primarily from the know-how developed from practice. However, this is increasingly being supplemented by both technical and operational expertise. The design and construction process will be divided into stages to enable analysis and choice to be managed. Once stages are determined, the analysis continues through an identification of the operations required to prepare and assemble the works and the choice of methods to be used, which determine the resources necessary. How many stages it is necessary to divide the building process into will depend on the novelty of the construction, and particularly the perceived technical risks. For productivity, repetitive operations may also be seen as worthy of a high level of analysis.

12 Manufacture, Assembly and Disassembly

Once we understand the issues, then we can move to the manufacturing phase. This is where everything – analysis and choice – comes together. This chapter considers an approach to the analysis of the production process involving both manufacture and assembly. It suggests that it is not just the concern of the production manager but needs to be understood as an integral part of choosing an appropriate construction solution. Starting from the need to ensure performance, it is necessary to visualise the construction in pieces and then at its various stages of partial assembly. The work required to complete each stage has to be seen as a series of operations achieved through chosen methods that determine the mix of resources necessary. The resources of materials and labour are supported by two categories of production equipment: temporary works and plant and machinery. The methods chosen must aim for a safe, economic and timely completion to an appropriate quality to ensure performance. When the analysis shows that these can be achieved, the suggested solution can be said to be buildable. The chapter concludes with an exploration of the nature of *in situ*, traditional and prefabrication production options.

12.1 Realisation of performance criteria

In considering the physical behaviour of construction the analysis has taken the completed building to be under operational conditions. The proposal represents a potential performance. Only the building can realise an actual performance. It must, therefore, be part of the evaluation to question whether it can be built at reasonable economic and environmental cost, in a reasonable time, in a safe manner, and achieve the performance potential as set out in the project brief. Failure to do so will result in a performance gap.

The purpose of the analysis in previous chapters has been to specify and detail the components, their materials, shape and dimensions, and their relative positions in the completed construction. Any analysis of production needs this information to consider methods and resources that might bring these requirements to reality. This does not mean that it is impossible to think about production until the solution is finalised. Even if the solution is at an early stage, with only broad ideas of possible materials and dimensions, the early indications of production options are established. It is possible to recognise production implications as soon as a physical form is emerging. This may then influence the choice of construction as the solutions become refined. In simple terms we should be asking if the assembly process will be undertaken entirely on the construction site, entirely offsite in a factory, or a mixture of the two.

Barry's Construction Technology: Analysis and Choice, Third Edition. Stephen Emmitt and Tony Bryan.

12.2 Visualising stages and sequences

Analysis of the behaviour of the final construction requires the ability to visualise the construction fully assembled and under working conditions. To consider manufacturing and assembly processes it is necessary to visualise the building in pieces, to be able to see the materials and components at various stages of manufacture, to see how each part may look when it arrives on the site, and how it must be prepared for incorporation into the final assembly. This then leads to the ability to visualise how the building will look at each stage of construction, when partially erected. This is necessary to analyse not only the methods and resources of production but also the behaviour of the partly assembled building to ensure it remains in a safe condition at each stage of production. The visualisation is made possible with BIMs, which have the capacity to model the assembly and disassembly sequence virtually. This helps to identify clashes and safety concerns before construction or disassembly commences.

The assembly and disassembly sequences must be visualised as a series of stages. It is then necessary to identify the operations that must be performed to safely move from one condition to the next. These operations can often be undertaken in a variety of ways, each with their own mix of resources that must be of sufficient quality and be available in sufficient quantity. Being able to visualise an appropriate method to produce the building safely, economically and to achieve performance is the criterion in declaring the construction to be buildable. The basis of the analysis of buildability is to be able to identify all the operations that are necessary to complete each stage of the construction. A very large range of operations is involved, from handling, preparing and assembling the parts, to technical and managerial control. Each operation must be identified, and a method chosen that will determine the resources necessary to achieve the next stage of production. It is necessary to visualise these stages, identify the operations required and decide on methods before it is possible to check that the resources can be obtained to complete the works.

Manufacture and assembly

The overall process of production can be characterised as taking processed materials, forming them into components and then assembling them to create the building. Among the primary decisions to be taken in this process is where these activities should take place. In traditional construction the components are created and assembled by craft skills. This involves cutting and fitting on the site, leading to site operations dominating the process. This is not a good description of some contemporary building. Components are made and assembled off site. Components arrive on the site as sub-assemblies, or prefabricates, made to size with no expectation of site adjustment. Not only does this mean that the production process now becomes more related to manufacturing but also site practice becomes focused on jointing and fixing processes to allow for adjustment and fit. In yet another process, *in situ* forming of components in their final position can be used, as, for example, in the reinforced concrete frame. Offsite manufacture may be adopted for the temporary works and reinforcement, while the site operations are focused on forming.

While some highly prefabricated systems have been developed, building is still likely to involve a mixture of traditional, prefabricated and *in situ* processes. It is, therefore, important to be clear about the balance between manufacture and assembly sequences both on and off the site. This is then related to the availability of trained operatives, technical and managerial staff with the skills necessary to achieve the quality required within any constraints of time. The aspects of these production options are discussed in more detail at the end of this chapter.

Identifying stages

The process of production involves taking processed materials, forming them into components and then assembling them to create the building. Production can be seen as a series of stages, known as activities, each defined by the creation of the next part of the building. Each one of these activities will take place in a sequence and will be associated with a series of operations that suggest methods and sets of resources that will bring the construction to a natural point at which inspection may be made prior to starting the next stage. Stages can be as large or as detailed as required, depending on the stage of thinking about the production process. Even at the design stage it may be necessary to think through some parts of the building in some detail, perhaps because of the innovative nature of the design. When the production process is being planned it will be broken down into monthly and daily programmes of work to enable the safe and efficient scheduling of work packages.

The reason for breaking down the construction into activities is to gain a sense of the methods and resources required, which will determine cost, time and safety aspects. It is difficult to lay down rules as to how big these stages should be as it depends on how novel the suggested solution is. In well-established solutions with well-known details, whole elements such as foundations, walls and roofs may be sufficient. If the solution is unfamiliar, care must be taken to think through often relatively small activities to be clear that the suggestion is buildable. When these divisions become very fine, they become part of the craft or trade, and they can be left to the individual undertaking the operations to question the buildability of a suggested solution.

One other type of activity that may be considered in some detail right down to the operational level is one that is repeated many times in the assembly of the whole building. Small savings in repetitive tasks and operations can bring high rewards in both cost and time without jeopardising performance if they are thought through at the time the solution is being evaluated. This type of analysis is important in evaluating details for highly prefabricated systems. More generally, the depth of analysis required is associated with the risks involved. Risks are normally associated with the safety, technical and financial aspects of a project. These are normally at their greatest with novel construction or where site conditions are complex, such as working in or near buildings that must remain in operation.

Safety is a prime reason why it is necessary to carry out a risk analysis. The danger to health and safety at any stage of construction can often be reduced by changes in materials and details. Mitigation of health and safety risks must be a constant aim in the choice of a solution as well as in the analysis of the production operations. Technical risk is associated with the possibility that the building will, through inappropriate production methods, not perform when completed or will suffer a premature failure either during construction or in the early years of its occupation. This is most often associated with not being able to achieve the required specification of materials, quality of workmanship or dimensional accuracy to ensure performance in the part(s) of the building to be completed subsequently.

Although both safety and technical risks carry a financial penalty, some risks directly affect budget expectations. Financial risk of this type can arise in many ways, often not just associated with the direct costs being more than anticipated. An example of this is financial penalties that arise from not meeting occupation dates. Risks to meeting occupation dates can be reviewed by identifying activities that are on the critical path. This defines the stages that form part of a chain of activities that represent the minimum length of time of the project. Delays in these activities directly affect the date on which the building can be handed over to the client. Planning techniques such as critical path analysis can be used to identify critical activities. Knowing in some detail the methods to be employed on these activities will lower the risk of running over time and incurring financial penalties, without jeopardising the performance of the occupied building.

12.3 Analysis of operations, methods and resources

Once decisions concerning the level of analysis required have been taken and an appropriate division of activities and a sequence identified, the starting and finishing conditions of each activity can be visualised. It is then necessary to identify the operations required to complete that stage of the work. From an understanding of the operations, the methods and their associated resources can be chosen for the stage under consideration. However, the methods and resources may be influenced by those required for other stages, to give continuity. Repetition of operations and continuity of resources have an influence on productivity and the time–cost–quality relationship.

Figure 12.1 shows a cross section through a concrete strip foundation that supports the cavity masonry wall. While this is typically what would appear on the construction drawings, there are other activities or stages in the construction that must be undertaken to achieve the design. These are normally associated with the production equipment discussed later in this chapter. In this example these would include the removal of the topsoil and the excavation and possible support of the trench. Figure 12.2 shows the sequence as the activities are shown on a production planning Gantt, or bar chart.

It is now possible to identify the operations associated with each stage in the construction of the foundation. This requires the introduction of a new vocabulary to visualise and describe the proposed construction. Activities or stages are described by reference to the parts

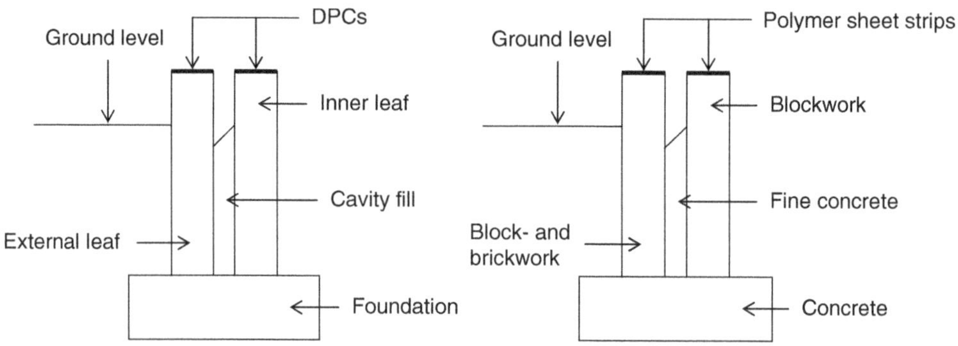

Figure 12.1 Strip foundation.

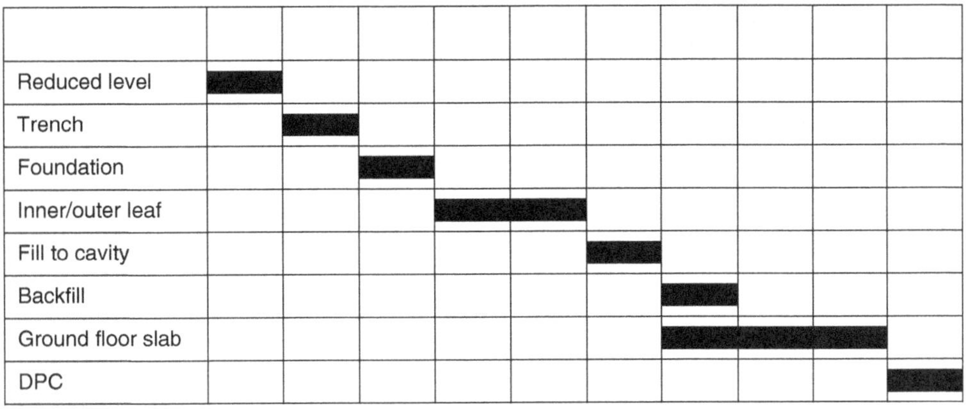

Figure 12.2 Strip foundations – stages and sequence of construction.

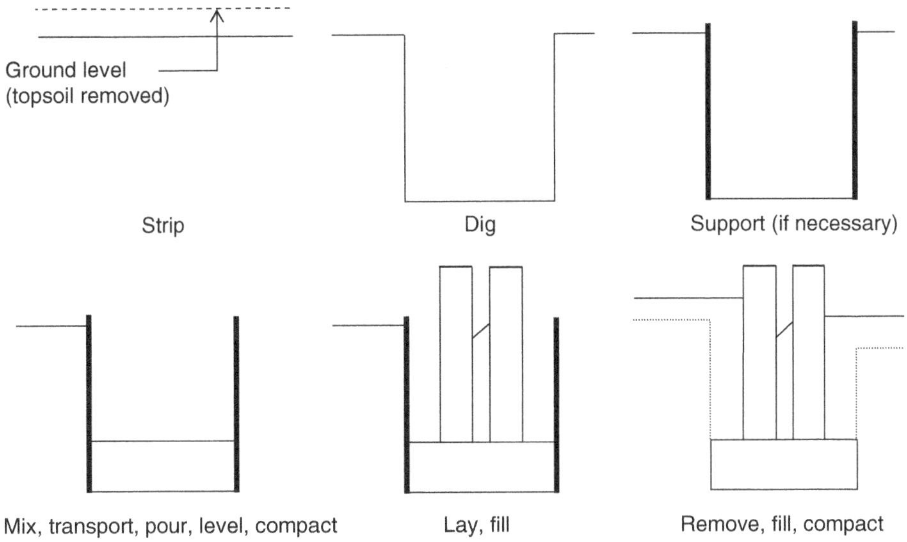

Figure 12.3 Strip foundations – operations associated with stages of construction.

of the construction to be created: foundations, wall, damp-proof course (DPC). Figure 12.3 suggests that operations are actions and therefore need active words (e.g. dig, lay, pour, transport, etc.) and illustrates how often the activities involve several operations. Trench would involve digging and may also require temporary support. Temporary works do not normally appear on construction drawings. Only an understanding of the production process allows the identification of the need for temporary structures and some ideas as to what methods may be appropriate and resources necessary to ensure safe and economic production. Forming the foundations will involve the chain of integrated operations associated with concreting: mix, transport, pour, level and compact. All these must be controlled to ensure that the quality of the concrete and the position of the concrete are within specified tolerances.

It is these operational actions that give clues to the available methods and the mix of resources that could be employed. For the excavation of the trench, hand methods may have to be employed in very confined spaces but are unlikely to be economic if access is available for machines to be used. If machines are used, the skills of the labour change to driver and bankspерson, whose job it is to guide the machine driver and ensure safe operations. Concreting is likely to be manual in the bottom of the relatively small trench, but options for the methods of transport will depend on the quantity, rate of delivery, and site conditions. For small works, wheelbarrows may be appropriate, but again on open sites machines for transporting the concrete will probably be more economic. Laying the DPC is likely to be part of normal bricklaying operations when the external walls are built.

Returning to Figure 12.2, the Gantt chart is used to show sequence and time. It is the method and the quantity of resources that determine the time an activity will take. This is clearly most important in production planning for the organisation of the appropriate resources in the right place at the right time and in the right quantity. However, even at the design stage, clients are interested in time as well as budgets, so time often becomes a key aspect in the choice of a construction solution.

It is now possible to see the range of resources that may be used in the manufacture and assembly of buildings. The direct resources involved will include the specified components and materials, the trained and skilled operatives required to complete the work and the production equipment necessary to assist in the production process. Management and technical support will also be necessary with the requisite expertise and experience. Indirect resources associated with corporate organisation and capital will also be necessary.

Resources may include the need for temporary works and safety measures to complete the work. In some cases these can form a significant part of the costs and time associated with the completion of the activity. This is the case with *in situ* concrete construction. Temporary works must be assembled and removed, and this involves a series of operations requiring resources. In a detailed analysis of the production process they may also be seen as activities. The possible methods and mix of resources are primarily determined by the choice of construction materials and detailing. Production cost and productivity have a baseline set by the construction chosen for the design. Careful choice of a production method and good management can only create efficiency gains. It is at the stage where construction solutions are chosen that basic cost and productivity are established.

12.4 Materials and labour

Fundamental to nearly all construction activity is the notion of labour (personnel) working on materials to bring them closer to their final state and to position them in the building assembly. It is a characteristic of industrialised societies that the materials will pass through several processes, worked by operatives with complementary skills and expertise. They may work in a variety of places, from sourcing the materials to final placing and finishing in the building. This division of work is reflected in the number of organisations that undertake this work and the levels of technical and managerial expertise that need to be applied. This in turn has implications for communication within the construction process, including those activities involved in the choice of construction.

Construction still relies on traditional skills, such as carpentry and bricklaying, although there is an increasing diversity of more specialised trades, some of which are related to offsite industrialised production. These are capable of a smaller range of operations, normally carried out within a well-defined stage of production associated with one part of the construction and aided by modern production tools and techniques.

Health and safety

While labour remains at the heart of the operations for construction, there is a major concern for the health and safety of individuals. Industrial operations carry with them specific risks. The analysis of the operations will reveal the hazards to which individuals will be exposed and the harm these hazards may cause them. When the chance of an incident causing harm has been established, the risks to individuals can be evaluated. A hazard that could cause great harm or a fatality, even if it is not very likely, should be considered as a high risk. More commonplace incidents may be considered as high risk even if they are unlikely to cause such severe harm. This type of analysis needs to be part of any evaluation carried out on a proposed production process. The evaluation of the risks may lead to changes in the choice of the construction specification and details as well as to the need for safety measures being incorporated into operational procedures.

Risk extends beyond the individual carrying out the operations. Any work may endanger others in the vicinity of the operation, either on the site or in its immediate environment. This includes not only the immediate danger of the accident but also exposure to both short-term and long-term health risks. Most of these risks are covered by regulations, and operating a safe place of work is a requirement of the law. The law places responsibilities on various parties in the design and construction of buildings.

The most effective way to reduce risk is to eliminate it by design. Once the design is established, the next approach is to change people's behaviour, and then to devise safe working procedures during the production process. Personal protection is considered the least desirable option. Although personal protection is still a major contributor to reducing accidents and long-term health problems in construction, design should aim to reduce the individual's reliance on personal clothing and equipment for their safety. The risk should be designed out. If that is not possible then measures should, be put in place to ensure that operations can be conducted safely.

The specification of construction materials and details can help to reduce this risk. Avoiding the specification of materials with known health risks from fibres, solvents, etc., can make a direct contribution to health and safety. Where the risk is very high, as was the case with asbestos, substitutes had to be found. With materials such as cement the risks are identifiable, but it is still specified, since alternatives are not easy to identify and protective measures, which are relatively simple, can be taken.

Detailing can help to determine where individuals must be standing when completing fixings and the size and weight of parts they will have to handle. Loads to be lifted should be limited and working platforms should be stable and allow posture to be maintained. Working at height, particularly with loose parts, increases the danger of falling, requiring more measures against accidents. Hazards exist when gaining access and egress from the working place, particularly if this includes carrying materials or components. Risks can be highest when operations are being carried out closely to each other, particularly if one is above the other. Other high-risk operations include working in confined spaces and lone working.

While this is not intended to be a comprehensive list of safety issues, the few examples given above are meant to indicate that risk evaluation comes from an analysis of the operations dictated by the choice of construction specification and details. Methods chosen must incorporate safety measures to create a safe working procedure. Real contributions to health and safety can be achieved in the choice of design and construction solutions.

12.5 Production equipment

Most building operations require operatives to handle and fit components by hand, assisted by powered tools and supported by a range of production equipment. The two major categories of this equipment are temporary works and plant and machinery. These have always been employed by constructors to make the work possible and safe. Typical examples of temporary works and plant and machinery are given in Table 12.1. This does not form part of the permanent construction and will be dismantled and removed once the permanent works are safe to use. Changing the construction choice changes the need for temporary works and plant and machinery.

In most cases the economy of these systems lies in their ability to be used many times in different locations. This means that any analysis of their useful life includes an analysis of the ease with which they can be dismantled and reassembled into similar, but often not identical, forms. This process of removal and reforming becomes a major cost in the life of many items of equipment. This may be as straightforward as transporting an item of plant between sites or the complete dismantling and reassembly, including retesting of falsework systems or a tower crane. This requirement of reuse has a profound effect on the approach that would be taken in their design, both in the standardisation of components of systems and, particularly, in the methods of jointing the components. Joints will be designed to facilitate quick assembly and dismantling. This can be seen in many temporary support systems such as scaffolding, which will need to be tested to ensure that the component parts remain reliable and safe despite frequent reuse. The reuse of production equipment means it is more prone to damage, misuse and unauthorised alteration, which creates the need for regular examination and retesting.

Table 12.1 Production equipment

Temporary works	Plant and machinery
Structural support	Materials handling
Partial erected construction	Excavation and earthworks
Forming	Concreting operations
Earthwork support	Manual operations
Existing building support	Power plant
Working platforms	
Site operations	
Access and egress	
Environmental enclosures	
Security	
Health and safety	
Environmental protection	

While both temporary works and plant and equipment have many characteristics in common, they have their origins in two distinct functions associated with the production process. Temporary works can be identified as fulfilling a primary function in that without them workers could not safely undertake the activities dictated by the construction solution chosen. Plant and machinery, however, have their origins in assisting in operations that were once carried out by hand. In this process of mechanisation, machines replace some labour (often associated with heavy, dirty or unsafe operations), although they normally generate new jobs. Much of what is now built could not be achieved by hand, as solutions are adopted on the understanding that certain types of plant and machinery are available.

Temporary works

There are systems that provide structural support until the permanent works can be used safely. Some forms of construction such as brick arches are constructed from components that only function as an entity once complete. This illustrates a fundamental question that must be asked about any construction solution: 'Will the components be self-supporting as the assembly progresses?' If they are not, there will be a need for temporary works to support the partially built structure. In the case of arches this is known as centring.

If the components are made from a material such as concrete that has no initial form, it will need a mould into which it can be poured. Concrete can be formed on the site in its final position, where it is known as *in situ* concrete, and the mould is known as falsework. This is usually made of timber or steel, and removed once the concrete has cured and is safe.

Concrete can also be formed in a mould offsite and then transported and fixed in its final position as a component; this is known as precast concrete. Since the materials from which the moulds are made are different from the material to be moulded, different trades may be involved. Thinking through the mould and support requirement for this type of construction at the time of design, and detailing for simplification and repetitions, can have a significant effect on the cost of the final construction. This has been seen in the design of *in situ* reinforced concrete slabs in the development from the beam and slab to flat slab designs, where the simplification of the formwork is the major source of cost saving.

Another major area of support that may be required is that of excavations. Because of the great variety of ground conditions and configurations of permanent works below ground, this is a major area of consideration. Where excavations are deep and/or near existing structures, earthwork support often involves a full engineering design. The need for earthwork support has implications for both cost and construction time. The choice is often left to the constructor, where considerations of sequence and difficulties of constructing permanent works around temporary structures must be weighed against the risk of ground movements during construction. However, the development of solutions such as diaphragm walling and contiguous piling, allowing the permanent works to be installed without traditional earthwork support methods, has made a major contribution to the time, cost and safety of the construction of basements, particularly in city centre developments.

When works above ground are adjacent to an existing building, or part of an existing building that must be retained, or alterations are being carried out, the structural support offered by the original building may have to be temporarily removed. The permanent works will have to reinstate this support, thus while the work is proceeding there will be a requirement for temporary support. This is normally known as shoring, and it will have to be designed by a structural engineer to ensure it can support the structure.

The temporary works identified above have all been to support part of the construction or the ground until the permanent works can safely take the loads. The final group of temporary support systems with a structural role is working platforms. These do not support the construction, but support the labour and materials associated with the operations as they are carried out. They are required until the operations of assembly are complete. Working platforms are most often scaffolds and must support the loads of the components and materials as well as the individuals themselves. Because of the direct use by individuals to carry out operations, health and safety becomes a major feature of the detailing of the scaffolding. It is important to consider not only the posture and ergonomics associated with carrying out the operations but also the possibilities of accidents.

Statistics show accidents from falling as a major source of injury and that most of these occur when work is being carried out at height, not only on temporary platforms but also from the permanent construction. Dangers from falling include individuals falling and materials or tools falling and injuring those below. This makes edge protection of particular significance. On temporary platforms such as scaffolds these must be incorporated in the design as handrails and toe boards. If the permanent works form the working platform, temporary edge protection will be required.

Detailing can often dictate where the operative should stand and what operations must be completed on or near to the edge of the building. For example, fixings for cladding can sometimes be arranged to be on the top of the slab or on the face of the edge of the slab. Fixings can involve drilling or the use of cast in fixing channels. While these have implications for tolerances and achieving fit, they also determine the options associated with the operations to be carried out at the edge of the building. Mobile working platforms are of particular use when operations must be carried out in a series of relatively isolated positions, at height, above a clear area, with an adequate surface for the equipment to move around safely.

Many systems will be associated with temporary access for the movement of materials, components and individuals to the place where the operations are to be carried out. Access (getting to) and egress (getting from) the place of work cover both horizontal and vertical movements. The type of construction will determine the range of loads and number of movements.

Analysis of these temporary works is at the heart of site organisation. Planning safe routes for the transfer of materials, areas of storage and methods of handling must be considered with aspects of temporary access. The loads to be carried, the routes to be taken, and the physical size of access requirements determine these routes. Cranes carrying materials through the air need loading and landing areas, while hoists require support from either permanent works or temporary scaffold structures. Vehicular access not only will need roads as well as hard standing for unloading, but also may need environmental control equipment to limit the spread of mud. Separating out pedestrian access from vehicular routes contributes to safety. The link here to plant and machinery is very strong. The examples above of cranes and hoists for materials handling indicate how levels of mechanisation determine the need for temporary access and egress structures.

While the idea of individuals having to move both to and from the working area is straightforward, components and materials only must be taken to the place of assembly. All operations create some level of waste. This may come from preparing materials, shaping components, cutting to fit or just from the packaging in which the components were delivered. Some estimates of the amount of this waste and the form it will take have to be considered when planning its recovery, recycling and reuse.

Temporary construction may have to be provided for environmental enclosure. Nearly all sites will require temporary office and storage buildings, the extent of which will be proportional to the value of the works. It is possible on some sites to prepare materials or even complete sub-assemblies in areas on site, but not in the final position in the building. These operations may be covered, creating workshops or small manufacturing areas. It may even be beneficial to provide cover for the assembly process if adverse weather is expected, although many assembly operations would be carried out under open site conditions.

Security works will be necessary at the boundary of the site to protect the works and to limit access of the public to the potential dangers of site operations. Further levels of security may be required for storage compounds to protect materials and components from theft and vandalism. Some items, such as bottled gases, pose safety hazards and these may carry a legal requirement for special secure storage conditions.

Although the temporary works so far identified will achieve most of the health and safety provision, some operations may need specific provision. Areas of danger such as edges of excavations, overhead power cables or holes in slabs will also need protection.

Many site operations may impact the natural environment. This may include the soil, air, watercourses, or the local wildlife balance. While this is best avoided by devising procedures that do not affect the environment, it may be necessary to provide specialist disposal and/or temporary screening or settling arrangements in some cases.

It has been suggested that some temporary works are necessary as support structures and some because of the needs of the overall site operations. In all cases they are influenced by the choice of construction. Analysing how a proposed solution must be manufactured and assembled includes identifying the temporary works that the suggestion will require.

Plant and machinery

The distinction has already been made that, while temporary works are fundamentally necessary to assemble buildings, plant and machinery have their origins in the mechanisation of manual operations. However, many buildings could not be built without machines as the design and choice of construction is made in the knowledge that the plant and machinery exist. An understanding of the levels of mechanisation that are available must be part of the analysis of the proposed solutions.

The most successful set of manual operations to be mechanised is materials handling. The size and weight of loads that can be lifted and carried have a direct influence on both the size of components and the potential size of the building. The range and scope of materials handling plant are very wide. They vary not only in the size of loads they can carry but also in how versatile many have become, being able to move a variety of materials and components around the site and into position.

The most obvious distinction is that between horizontal and vertical movement. Dumpers, trucks and lorries move horizontally, hoists and winches vertically. Cranes, forklifts, pumps and drones can achieve both horizontal and vertical movement. All this plant has an operational envelope defining the limits of movement in the horizontal and vertical planes that must be matched to the site access and the form of construction chosen.

Excavation and earth moving is a specific site activity that has developed a range of plant for specialised tasks. The action of digging involves additional forces over and above the action of picking and carrying the load. Machines range from the versatile slew machines with a range of buckets to specialist trenching machines. Pumps are used to mechanise the manual operation of bailing out flooded excavations. There is a range of lorry and dumper transport options, again dependent on the size of operations.

Another site activity that involves a range of operations that have been mechanised is concreting. While most of the mechanisation is materials handling, other operations are associated with preparation and final assembly. For all but the smallest amounts of concrete mixing will be mechanised, ranging from small hand-fed mixers to fully automated batching plant. Moving the mixed concrete to the formwork can use a variety of materials handling plant, depending on the site conditions and the stage of construction. Once in the formwork the compacting and even the spreading and levelling operations may be mechanised. One concreting activity that has become even more mechanised is the laying of large areas of concrete ground-supported floors. Not only is the plant highly specialised in placing, spreading, levelling and finishing, but also it incorporates elements of automation.

Many operations remain essentially hand operations but are assisted by hand and bench tools. Cutting, drilling, screwing and spraying are just some of the hand operations that are now commonly undertaken with power tools. Many of the smaller versions of these tools can be battery operated, but for larger, more powerful applications a power supply must be available. Electricity and compressed air are the two main sources of power supply for small tools. This may require the use of generators or compressors as on-site power plants.

Most plant is either handheld (some simple machines such as winches are even hand-operated) or controlled by a driver. There have been some attempts to automate the process with driverless plant using computer control. While this has found limited application on construction sites, it is increasingly used in offsite factory production. The use of computer-controlled technology in manufacture is leading to systems that link electronically produced design information with the manufacturing facility to automate the fabrication process.

Understanding the process and possibilities of mechanisation, digitalisation and production identifies the relationship of the construction solution to the production process. Production analysis becomes a necessary part of the evaluation of a suggested construction solution.

12.6 Production options

The production process can be characterised as taking processed materials, forming them into components, and assembling them to create buildings. This implies a sequence of operations each with an associated method that defines the resources required to complete the work. This series of activities associated with processing, forming and assembling is greatly influenced by the detailing and specification chosen to realise the design.

To carry out an analysis of this production process the construction must be visualised in pieces and envisage each part at each stage of its manufacture and assembly. Visualising the components and the building partially formed at each stage of production and the operations to be performed to bring it closer to the finished building is as important as being able to see the building completed and responding to operational conditions. This can be done by using a simple storyboard type of graphic for simple building operations. More commonly the process of assembly and disassembly is created within a BIM to explore the feasibility and safety of operations before work is commenced. This can result in cost and time savings as well as highlight any concerns over buildability or safety.

Very early in the design stages of a building the decisions that are taken influence the nature of production activities, not only the chain of events necessary to produce the building but also where these events may take place. There are broadly three places where the processing, forming and assembly take place:

- In position
- On the site
- Offsite

The final stage of assembly must take place in position but there is a question as to how much of the processing, forming and assembly can be completed either adjacent to the emerging building on site or offsite under factory conditions. This choice of where operations take place is often closely related to the choice of materials and the trades and processes associated with their manufacturing and assembly. For a traditional material such as timber the skills and know-how have been developed to work the material both on site and in the workshop. Initially this may have been associated with carpentry and joinery, but modern manufacturing technologies have allowed development not only in the scale of forming and assembly in the workshop but also in the processing of timber to form boards and laminates that have greatly expanded the possibilities to employ timber in a wider range of building applications and design concepts.

The possibilities to employ steel in building came with the industrial scale of processing, and this remains true today. The forming of steel is also an industrial and workshop activity with most steel items being delivered to site in component form ready for assembly. While it is possible to see some forming, cutting and drilling take place on the site with wire, bar, or sheet materials, this is less appropriate with heavier sections. The need for such site operations may warrant establishing a temporary workshop to complete the operations.

Concrete as a material that needs forming in a mould has developed in two major forms: *in situ* and precast. In both situations it is important to consider the jointing and fixings, where tolerances must be accommodated and fit achieved. Where moulding should take place will be dependent on factors such as size of component, technical aspects of the materials and specification, number and type of components involved as well as lifting and transport implications. Moulding may be completed on the site to minimise transport but will need some workshop type of operation to be established in, for example, tilt-up construction. It may be completed in a factory dedicated to the production of these components and then transported to the site.

These manufacturing and assembly options are determined very early in the design process, not only in the choice of materials but also in the detailing. Starting with the choice of dimensions and the repetitive nature of components, this can be extended to the detailing of joints and fixings to determine the boundaries of assembly, disassembly and prefabrication. Prefabrication is the assembly of several components, possibly of different materials, in factory or workshop conditions to be lifted into position later. Again, this can be achieved in on-site temporary assembly shops best placed close enough for direct lifting into position but can be produced in factories away from the site and transported in their preassembled condition.

This simple analysis of materials and their traditions in processing, forming and assembly and the discussion of where each of these three key processes takes place give a picture of three broad types of production options:

- *In situ*
- Traditional (incorporating factory-produced components)
- Prefabrication and system building

While there are distinctive features of each of these options, it is important to be aware that they often exist on a continuum where it is difficult to identify the move from one approach to the next. It is also possible for different elements of the building to use different approaches. If this is proposed, the inherent deviations and tolerances in each of these approaches need to be resolved in the junctions between different approaches.

In situ processes take essentially formless materials such as soil or concrete and create components in position. The whole process of forming size and shape takes place in position. Only the preparation of the material takes place prior to the forming of the components. The skill and know-how associated with the temporary works often plays a major part in the success of this type of production as this determines not only shape but also dimensions including achieving tolerances and fit.

Traditional construction is based on operatives who work on site, where the individual has the skill, often working with one material to shape and/or fit the components, and then fix and joint the components into position. Originally the skills of these operatives would include the ability to carry out many of the shaping and forming processes for the components on site. Increasingly, however, the components are made off site and, while traditional trades are still employed, the nature of their work on site is to fix and joint the components. Components are not normally designed to be cut on the site. In fact any need to cut and fit the components would be considered a failure of control of tolerances and fit. Often these components may be built in, gaining fit by cutting the surrounding, more traditional construction. Typical of this change would be the work of carpenters with the use of gang-nailed roof trusses replacing the timbers of the cut roof. However, with more components being used there is less opportunity to build in, so fit must be accommodated at the joint or some cutting to fit designed into the component.

The use of components in traditional construction has led to the development of what are often seen as specialist systems, normally erected by operatives who do not have the wide range of versatile skills of the traditional trades. Specialist operatives have skills associated with a small range of standard sections and components. There is often still a need to cut and fit some of the standard sections, but these are often then connected or supplemented with a range of components. This can be seen in many of the internal finishing systems such as metal stud partitioning (dry lining) or suspended ceilings.

The development of manufactured components has created the opportunity to design significant parts or even whole buildings from standardised components. Often known as system building, components are manufactured in a limited range of sizes to a dimensional framework with standardised jointing and fixings for site assembly. In these systems, cutting and fitting on site will be minimised in the design and should be eliminated for the pre-sized components on site. Fit now must be accommodated in the jointing, often with adjustments being designed into the fixings. This represents a change from the traditional site processes even if they do include the use of many components.

Many components delivered to the site will be manufactured and assembled in a factory. Many parts may be fixed together before being delivered to site, for example the door hung in the frame with lock or latch fitted in door sets. Prefabrication is the term given to an assembly of components that are put together in a factory or assembled on site prior to being positioned in the building. Prefabrication can range from the production of a basic structure onto which enclosure, finishes and services can be attached on the site to the prefabrication of whole sections of a building complete with cladding and services to be lifted into position on site. These options are discussed in more detail below.

While the possibility of on-site manufacture and prefabrication exists, the current levels of infrastructure both in manufacturing capacity and transport means that prefabrication can now be achieved in factory conditions remote from the site. Initially components, systems and prefabricates had to standardise shape and dimensions for production efficiency and to lower cost, but increasingly flexible manufacturing methods mean that bespoke sizes and shapes can be produced economically. An example of this is the production of timber or steel wall frame panels for single one-off house designs. This can help to reduce waste and improve the environmental impact of the chosen construction method.

Time, cost and quality

The need to consider a particular production option for a project can be associated with resource availability, often either materials or skill shortage, to achieve construction against the criteria of time, cost and quality. It is a mistake to think that there is direct relationship between some of these criteria and the production options. The production options for any one project will depend on the context and prevailing resource base, including the design skills currently available. It may be better to adopt different basic production approaches for different parts of the building. However, as pointed out above, this needs to be done with care to match deviations both induced and inherent, to ensure manufacturing fit and accommodate long-term movement respectively. It puts a specific focus on the interfaces or junctions: the actual joints and fixings between elements of the construction with different production approaches.

It is important not to judge speed based only on the time on the site. This may lead to the impression that prefabrication is the choice for speed. The timescale for the client (who may have to pay a premium for speed) is from commissioning the design to hand over for occupation. This includes all the design time, the pre-production lead times, the factory production time and the site operations. If a system, fully thought through with all the components, joints and fixings standardised ready for ordering into a variety of building configurations is available, some of this time can be reduced. This assumes that the scale and context of the building is suited to the pre-designed system. If the design is to be specified and detailed for bespoke prefabrication, the design, procurement and manufacturing time needs to be anticipated, as well as the time on site. For many traditional designs the materials and expertise are widely available and can often be procured and mobilised relatively quickly, even though the time on site may be longer but still giving the client an earliest occupation date.

To produce components and prefabricated units requires capital investment in factory and manufacturing capacity. This achieves economy only if steady demand exists to maintain manufacture at a sufficiently high level for the capital to be recouped. While this can be achieved in many components such as lintels for cavity brickwork, which are specified in a range of types and scale of building, it can make more bespoke and complex prefabricates expensive. The need for a demand for a limited range of standardised components is less true in the more flexible manufacturing plants but the machinery to give this flexibility, often based on computer numeric control (CNC), is expensive but can accommodate changes in the detailing and specification, particularly of dimensions, very easily. This gives the opportunity to provide one-off designs for individual clients. The economic costs and the need for a continual market demand are why many prefabricated systems have failed to become established.

Capital costs can be reduced if the prefabricates are detailed as traditional construction. This may allow a flow line approach with traditional trades working in a workshop environment. This has inherent improvements in working conditions, including health and safety, with the potential for reduced time and increased quality.

Quality is the third reason to carefully consider the production options. Quality, like time, can be a justification for increased cost if it is of value to the client. All buildings need to be formed and assembled with care and levels of skill and knowledge and quality of materials appropriate to the design. Each production option, with the material chosen, has the capacity for quality both in finish and dimensional accuracy, important in realising the design concept. There will be associated with every design a level of dimensional and workmanship quality that if not achieved will not realise the design intention. It becomes part of the analysis of a technical solution to check if the production process implied in the specification and detailing can deliver the quality required.

Sustainability and environment

Sustainability analysis will be associated with production options. As buildings produce less carbon in operation, the CO_2 release associated with the materials in embodied energy and carbon becomes more significant, and this includes the production process. Embodied carbon in materials is mainly associated with processing the material and then in the forming of the components and the assembly process. In the full life cycle analysis all these stages must be considered, including dismantling and recycling options. Figures for embodied energy and carbon materials are normally quoted from cradle-to-gate, which normally means from extraction to the site gate. Increasingly energy and carbon figures are associated with materials formed into components to help to refine the analysis, but carbon released is also a concern of transport and assembly processes. This leads to notions of local supply of materials and components, but this must be extended to the full journey of the materials and components if prefabrication even from local suppliers is to be considered. *In situ* and the other trade-based processes require the delivery of materials and components to site and then distribution around the site. On-site (or even local) fabrication may prove effective in limiting carbon. Offsite fabrication process may require more transport and more machinery but may mean less waste. Thus the analysis and choice may be complex when it comes to trying to establish the option with the least environmental impact.

Waste is a key issue for the environment. In addition to concerns for embodied energy, processing will produce solid, liquid and gaseous waste that can pollute our water and air. This type of waste is normally less in the forming and assembly phases and more associated with materials processing. However, many factory and site processes produce waste, and without design and management controls the waste may not be reduced or recycled. It is a challenge for site operations to reduce waste and to have procedures in place to separate and stream waste disposal to reduce landfill and promote recycling. Once again, this can be mitigated by careful choices at the design, detailing and specification stages to design out waste, or at least mitigate the impact.

Demolition operations have a specific waste problem which has led to the development of many technologies and procedures to deal with demolition waste streams. Brick and concrete can be crushed on the site to produce fills, while steel can be collected for recycling and forming into new products. Timber can be selected for reuse or sent to be processed into timber-based composites or for fuel. It should now be a conscious design decision to use materials that can be reused or recycled at the end of the building's life. This can be further aided by considering dismantling in the detailing and choice of non-destructive fixings. Ease of disassembly can greatly improve the chance of reuse of a component or material.

Factory operations can manage material waste successfully as they have a predictable supply of waste that can systematically be reduced or put to another purpose. In the production of timber components waste can be processed into boards or larger structural sections, processed for insulation, or even burnt to provide energy and heat for the factory.

Both the choice of materials and the production process have implication for energy, carbon, waste, transport and the opportunity to procure from sustainable sources. To these can be added issues of water used in the processing of materials and fair trade in the overall sustainability argument. Both the specification and procurement processes can help in responsible and ethical sourcing to meet sustainability objectives. This means that there is a need to evaluate the materials and components not just one with another, but from alternative sources of supply. Rating and certification schemes such as BREEAM and LEED can help in this evaluation, but the schemes vary in their complexity, the factors that they consider and the amount of explicit information they give. These are useful, for example the scheme to certify that timber has been obtained from sustainable or renewable sources. However, for a major supply contract involving a significant proportion of the works or where particular pollution, transport or scarce or over-exploited resources are involved, individual evaluations based on the above analysis may be necessary.

Production options – design decisions

The possibilities for production options are significantly affected by the choice of technical solutions at the design stage. Each of the technical choices made at the design stage comes with a set of production options. The production options are often very limited, being closely associated with the development of the technical solution.

Well-established (robust) detailing implies a set of resources and methods that have been developed to make those details work. These are normally associated with *in situ* or on-site trade operations and will achieve required performance levels only as long as they are produced in an appropriate way, often a way that has been established over time along with the refining of the details themselves. New, untried detailing may imply a change in the production process and any desired change in the production process will imply a re-evaluation of the detailing.

The need to consider new technical options to achieve new performance requirements based on the demands of society and of clients is constantly changing. This is in addition to the pressure from the industry to achieve economies and efficiencies in production and this can often best be achieved by modifying details and specification. All designs must ensure that an appropriate production process is viable to realise the building. Some may even include exploiting a particular production option to express a design concept.

Prefabrication and system building

Prefabrication and system building employ factory made items that need no cutting on site, often made in a limited number of coordinated sizes to control production costs. Therefore, the first decision to be made for systems building and/or prefabrication is the identification of a dimensional framework. Dimensional frameworks and repetition bring economies and efficiencies to site-based trades. However, in prefabrication a dimensional framework is necessary to limit the number of pieces to be manufactured yet give maximum freedom to the design of space for the efficient use of the building. Often identified as a module size, it is chosen to determine the size of prefabricates to give maximum flexibility in design. The most used module is 300 mm leading to larger prefabricated units being based on 600 mm, 1200 mm or even larger units. This not only gives economic division of internal space but must provide units that are easy to handle. As indicated above, the introduction of flexible manufacturing methods has reduced the need for this standardisation but some dimensional discipline will lead to economies in the detailing and production process.

In addition to the choice of a dimensional framework, decisions must be taken on the extent of prefabrication for off- and on-site processes. High levels of prefabrication involving finishes and services off site reduce the time on site, however, it increases the chance of damage and can make the design of joints more challenging because they may be visible between prefabricated sections. While it is a principle of prefabrication that cutting and fitting on site should not be necessary, the greater the level of assembly achieved off site makes this even more imperative as any modification will almost certainly affect appearance as well as potentially decrease performance.

The simplest level of prefabrication provides only a basic structure on which the other elements can be fixed. While this can be the loadbearing elements of the building carrying the dead and imposed loads, it can be a sub-system of panels or pods that will carry the enclosure elements such as insulation and finishes. For example, precast concrete panels can be employed to form a loadbearing structure or cast as cladding panels to stand or hang on a building frame. Timber-framed or steel-framed panels can also be used as loadbearing or infill construction. These framed panels are normally delivered to site as empty panels with boarding on one side ready for site processes of insulation and services carcassing prior to boarding the internal surface and cladding the outside, all as site-based trades. However, panels may be delivered with both sides boarded, and thermal insulation installed ready for finishes to be applied on site.

Another approach is to develop the system of building components made off site and delivered to be assembled into the building but with standardised joint and fixings. In many of these systems the components can only be used with other components in the system.

These can vary in scope from single-element systems such as building frames to systems that create whole buildings. The dimensional ranges of the components and the detailing of the joints and fixings become particular to the system. These may be developed and procured through commercial companies or through consortia of clients who have demands for similar buildings. Where buildings systems are developed, they may be limited to certain phases of the building, for example to watertight providing a dry envelope, or may include all components including internal partitions and fittings.

Rather than try to develop a system that can be applied to many buildings it is possible to design a bespoke system of components for a single project. This system approach combines the offsite manufacture of components with on-site assembly. The on-site assembly is now not dependent on traditional trade processes, as it should eliminate cutting and shaping and provide simplified jointing that, while it may be carried out by operatives with a traditional trade, becomes an assembly process. This will limit the number of site operations, reduce the range of tools required and give repetition that, when learned by site operatives, should speed up site production. Speed on site can also be claimed by the commercial and consortia systems. However, they can benefit from the design and detailing investment over many buildings as well as economies from component manufacturers that have some expectation of repeating orders from several projects. The bespoke system may have an increased design time and require one-off negotiations with manufacturers, but this may be deemed worthwhile for the nature of the building or the design and build quality.

Another level of prefabrication is to complete a significant amount of the assembly of enclosure, finishes or services components on the basic structural unit in a factory before delivery to site. With more complete units being delivered to site, erection time can be minimised, particularly if joint design is focused on achieving fit and simplifying site operations. It may put more emphasis on tolerances and fit and leaves the construction more vulnerable to damage.

As these issues become more of a concern it leads to the notion of volumetric prefabrication. This involves replacing panels and flat pack deliveries with whole rooms or sections of building, including floors and walls as pods. While the outside of the pod may need to be clad later, it leaves the inside less vulnerable and therefore finishes, services and even fixtures and fittings can be completed in the factory within the pod.

Potential for prefabrication can be identified if the building programme is examined to identify either part of the building where early completion is helpful or lengthy activities on the critical path, particularly around congested areas, take time to complete. Two such areas are staircases and highly serviced areas such as bathrooms and kitchens. These do, of course, need to be repeated many times to make the setting up of factory prefabrication economically viable.

In the past, systems and prefabrication have been associated with building types such as houses, schools and hotels, at a time when demand for these types of building has been high. It has also been successfully applied to high-quality, one-off development. It has also proved helpful to consider elements or parts of the building to be prefabricated to speed up overall project times. It is necessary to evaluate each building in its design and resource context to determine the merits of a particular production option. However, it is also necessary to be clear which production option is being implied by the design to ensure coherence in the detailing. The junction where different production options meet needs to be designed with particular care.

12.7 Knowledge and expertise for the analysis of the process

Carrying out this analysis of manufacturing and assembly requires knowledge and expertise gained from practice. At the most fundamental level this knowledge is derived from craft and operative practice with modern methods, now including use of plant and machinery from hand tools to high levels of automation. The basis of the understanding of the ways in which basic operations can be achieved comes from the know-how of those who work the materials and components and complete the assembly tasks. These are the people who know what can be achieved within the deviations demanded by the design and the time this will take. They will also have a view on temporary works and safe working procedures.

With modern specifications of materials this know-how must be supported by technical knowledge. When new materials are introduced, the key aspects of the operations that control their properties, and therefore performance, may not be immediately apparent from previous know-how. This must be worked out through the technical knowledge, although with time this may become part of common practice. In some cases the technical support becomes part of common practice. An example of this is the determination of concrete strength to allow the decision to remove the temporary works.

This is then overlaid with knowledge and expertise gained by those who organise and manage the production process. They can take an overview of the whole production process. They must see all the stages or activities that must be carried out. Management will gain a view on sequence, temporary works, factory or site organisation and safe working practice. They have a perspective concerning overall productivity, cost and time. They then need to be informed by, and understand, the operative and technical knowledge to ensure it is integrated into the plans and programmes that they provide.

In the same way that the choice of the construction is guided by an overall design but must be solved in the choice of specification and details, so production must be guided by overall programmes but must be solved in choice of methods and resources. These require a different knowledge base and approach to the analysis, but it does not mean that these two creative acts should be carried out independently. Overall productivity is governed as much by the design as the efficiency of the production operations. Overall performance is governed as much by methods of production as by the materials and details of construction.

PART II Choice

Applying the Framework: Chapter 13 AT A GLANCE

What? This text has set out to explain a basis for the technical and rational aspects of the process of analysis and choice. The narrative maintains that it is only through this understanding that creative and innovative ideas can be put into practice. To continue to apply this approach the chapters that follow are written from the perspective of a case study based on a small domestic dwelling in the UK.

Why? We need examples to help question the way in which we design and build to realise a better balance between the built environment and the natural environment. The examples are not put forward as how to do it, rather they are provided as a means of raising questions about how we build. The intention is to challenge the reader to think holistically and use first principles to explore innovative solutions to common challenges. This is supported with reflective exercises at the end of the chapters.

When? The framework identifies the areas to be considered but does not suggest the sequence in which the areas of analysis should be undertaken. The nature of each project will determine the key choices, with no single choice being confirmed until all areas have been considered, at least in outline. Thus, the when? question is contextual; however, we do urge readers to consider environmental impact in the analysis and choice from the start and throughout the analysis and choice process.

How? Analysis and choice are collaborative endeavours. The examples given in the chapters that follow can be addressed individually or as a small group exercise. In practice these and similar questions would be addressed collectively by the design team, working closely with the client, together with those responsible for providing materials, components, plant and equipment. The questions may also result in the need to explore appropriate digital technologies to assist with the analysis and inform the choices.

13 Applying the Framework

This text has set out to explain a basis for the technical and rational aspects of the process and maintains that it is only through this understanding that creative and innovative ideas can be put into practice. To continue to apply this approach the chapters that follow are written from the perspective of some aspects of construction for a small house in the UK. This chapter introduces the context in which these choices are being made, the nature of the environments, physical and social and identifies the aspects of design, appearance and the resource base that will condition choices.

13.1 The need for an integrated approach

Part 1 suggests that proposed construction solutions must be evaluated through several areas of analysis. There is a danger that this leads to the view that each area of analysis can be considered independently, and they can be carried out in a specific order to make the final choice. This view is not correct, and it is hoped that this text has not given this impression.

The first few chapters of the book are designed to give a holistic feel to the process of choice before the detail of analysis of construction is introduced. Each project has its own balance of priorities set by the client and its own context that determine the appropriate approach. The emphasis is on evaluation as a holistic process, which is not satisfactorily complete until all areas of analysis show acceptable predictions. While each analysis is largely independent, it is always being undertaken in a growing understanding of the outcomes of the evaluation in relation to a building as one entity. It is also being undertaken as a collective effort, with individuals working towards a common goal. This is facilitated by digital platforms and digital tools. Initial proposals need to be provisional until the outcomes of each analysis have become clear. This leads to the need to have techniques to establish approximate solutions for each analysis to see whether an overall solution is possible, as well as detailed design information to make the final choice. It also requires a managerial framework in which decisions can be reviewed and approved (signed off) prior to proceeding to the next stage or phase of the project. This requires a sequential process that is linked to the time parameters for the project, and which allows for iteration in analysis and choice. The project schedule and sign-off events (verification gateways) will be sequential, but the decisions will be iterative. This is why scheduled sign-off events allow for the possibility that not all decisions will have been resolved fully

Barry's Construction Technology: Analysis and Choice, Third Edition. Stephen Emmitt and Tony Bryan.

but are resolved to such an extent that it is possible to progress to the next stage with certainty that the issues can be resolved in a timely manner. Sometimes it may be necessary to return to the underlying principles to reconsider the choice if it is questioned at a sign-off event.

This is one reason why this text is based on a framework, not a sequence. The process has been thought of as being dynamic, open and iterative, moving towards a solution but without prescribing the route. The choice must be made within a framework that indicates the range and scope of issues that must be considered. A framework can be used to establish the complementary and competing factors that will determine the success of the building in technical and social terms. It is difficult to reproduce this process in the pages of a book. It is, by its very nature, one that must be experienced to fully understand how, what sounds like a chaotic process, is controlled and rational while also allowing space for creative choices. In practice a variety of project management and design management tools will be employed to manage these interactions and encourage informed decision making.

13.2 Individual and collective choices

Most decisions will be taken in a collaborative environment, with analysis of possible and preferred choices being influenced by other project stakeholders. The way individuals interact is beyond the scope of this text, however, it must be recognised that individuals cannot make decisions in isolation. At the simplest of levels architects will need input from structural and services engineers; contractors will require interaction with specialist subcontractors and suppliers; and facilities managers will require input from service suppliers and building occupants, etc. Decisions will be informed by available knowledge, shaped using environmental assessment tools such as BREEAM and LEED, guided by computer software modelling and simulation programmes, and made within prevailing legislative, cultural and social norms. This can be a complex process requiring clear communication and clearly defined responsibilities for making decisions. The use of collaborative information technologies, such as BIMs and digital project platforms, enables information sharing and collective decision making in real time. The challenge is for project stakeholders to have a shared vision for the building and for everyone to be committed to reducing the environmental impact of the construction process and the building in occupation. Viewing the building as an entire system in a whole life context, and as a closed loop, will influence the choice of solutions that have a positive impact on the environment.

We also need to recognise that our decisions, both individual and collective, are constrained by the time available to design and construct the building, and the budget (available finances). Time is a limited resource and one way of coping with the limited time available is to rely on solutions used previously (precedents and typical details). These are solutions that are known not to fail, i.e. they perform as expected. When we look at taking a more creative and innovative approach, for example questioning why we design foundations, cavity walls and windows and doors in the way we do, it will require time to go through the analysis and choice loop. And if we are going to improve the environmental impact of buildings, and turn a negative impact into a positive impact, we need to take more time to reconsider what we do and why rather than rely only on familiar solutions.

13.3 The basis of the case studies

The case study chapters that follow are focused on a simple domestic dwelling, a single storey house in single occupation dwelling in the UK. We can assume that the house will be occupied by a family, and the family needs and aspirations will change over time. This text assumes that these changes will not significantly affect the construction of what these families are seeking, although there may be future changes in the performance requirements of the family that may lead to alterations and retrofitting.

While this description of a house will allow a broad understanding of the needs of the user, it says nothing about the resource base, or industrial structures required to design and build houses. The two ideas used so far to define this case study - family and industry (demand and supply) - have set much of the scope of choice. The depth of understanding that is necessary, of each of these ideas, should not be underestimated to appreciate why certain construction solutions will be chosen. This was the thrust of some of the earlier chapters of this book, where it was suggested that it is necessary to understand something of the physical and social environments within which choices are made. This will include an appreciation of house design and the resource base available to the housebuilding industry.

Physical and social environment

The case study house is set in an advanced industrial society on a highly populated island in a temperate climate. The social values include the distribution of private wealth with equity and respect for all members of society, with a growing awareness of environmental impact of our actions. This gives a high level of demand for housing, with limited land and high prices on the open market, together with a need for social housing. These few facts start to create a picture of the context in which the choices will be made.

Advanced industrial societies often have a great deal of legislation and regulation. We can take advantage of factory production and transport of component and prefabricated sections of buildings from almost anywhere in the world. We are, however, having a huge impact on the environment. The need for housing on our highly populated island means that we are increasingly building on previously developed land, brownfield sites and have a problem in dealing with our domestic and industrial waste. Our concerns for equity and social justice, with a fair standard of living for all, set relatively high performance levels for construction. The temperate climate defines our weathering problems and indicates that water penetration and winter heating are major concerns for house construction. There are now concerns that the levels of human activity we wish to sustain are causing environmental changes that threaten our ability to maintain this progress. Environmental concerns must take their place with economic and social considerations to ensure sustainable development. The more significant aspects have been introduced in the previous chapters and the reader should become increasingly aware of the importance of these factors for analysis and choice.

The resource base

With high levels of housebuilding activity the supply of materials and factory produced components is good, but there is a shortage of site labour and the traditional trade skills. To meet demand attention has been paid to designing for prefabrication, including modular and volumetric systems to reduce reliance on site operations. Machinery is readily available for materials handling. The introduction of cordless hand tools has influenced fixing and jointing processes, often now the focus of the site assembly processes. Money for housing is readily available. The main pressure is on the use of resources in a more responsive and sustainable way.

Design and appearance

The basic design and appearance of the house is well established. The pattern of living is sufficiently common for the major rooms of living room, bedroom, kitchen and bathroom to be easily identifiable, with some circulation space including hallways and staircases. While there is variety in the number and layout of rooms, the size of the rooms has generally been reducing, particularly in the volume housing market where the building footprint and plot size is reduced. Externally the brick house is still a dominant form. Rendered finishes may be used, although they still maintain the sense of masonry construction. Roofs are pitched with tiles or slates to give the characteristic shape to most bungalows and houses. There is then a variety of window forms, including bays and oriels often to gain variety in housing design. It is not possible to generalise on the designs for the individual client, where expression in materials, windows and doors can have a major impact on the building's appearance.

Internally finishes are plain but need to accommodate a variety of building services. Energy saving has featured in the regulations over recent years and has had a large influence on detailing. In the single dwelling fire resistance is only nominal within the dwelling, but walls between dwellings (party walls) and the roof finishes that are continuous between dwellings are subject to more stringent regulation. New regulations on soundproofing and air infiltration with the possibility of the need for performance testing and the introduction of robust detailing are also changing the construction of and specification for houses. As we move towards airtight homes there is the need to use ventilation to maintain a healthy interior environment. Environmental concerns are affecting the balance between passive and active technologies as well as the choice of the components, materials and detailing.

Common form

The conditions established for the case study have given rise to the common form of a house. With the overall form established it is possible to take an elemental approach to floors, roofs, walls, foundations and services. Each element contributes a major group of functions and is easily recognised by the basic construction solutions. However, the question arises as to what order they should be considered in. It has already been established that no one element can be fully selected without some reference to the probable form of the others. In this case the elements that can be discussed knowing least detail of the other elements are the floors and the roof. These transfer loads to the walls, and so it is necessary to discuss these before walls can be fully investigated. It is then possible to complete the structure by looking at foundations where the loads from a house are relatively low and the site conditions are unlikely to dominate the choices for the superstructure. Finishes will be discussed with each element to cover all the analysis of the passive construction. This leaves the active services, although aspects of them will have inevitably been introduced by their need to be integrated with the other elements.

13.4 The case studies

Typical construction details and their description can be found in construction technology books, such as *Barry's Introduction to Construction of Buildings* and *Barry's Advanced Construction of Buildings*. Therefore, the chapters that follow do not deal with all the typical construction elements, instead they address some common issues in domestic construction. The aim is to question what we do and ask whether there may be a better solution to the commonly adopted solutions.

Chapter 14 considers foundations and asks the question: is it necessary to dig a hole and fill it full of concrete and steel reinforcement? Are there other solutions that are more environmentally friendly? While the answer will be influenced by the ground conditions for a project, exploring other possibilities is one way of confronting the use of materials and challenging the way we usually do things. Concrete foundations may be the most appropriate solution, but unless we challenge the familiar and established solutions it is difficult to say for certain.

Once the foundations have been considered we can then look at the walls and the openings within them. Chapter 15 investigates a wall typology that has become somewhat of a paradox as it has evolved. In most cases the cavity wall is no longer a wall with an air space (cavity). The 'cavity wall' now resembles more a sandwich or composite construction because the cavity has been replaced with thermal insulating material. There is no cavity; instead there are layers of different materials; all of which behave in a different manner when subjected to changes of temperature and moisture. This raises questions about the need for a cavity wall and what has become an overly complex construction. What other typologies could deliver a similar or better level of performance in a simpler and more environmentally friendly manner?

Openings in walls have long been a challenge for designers trying to match the thermal performance of the wall with that of the windows and doors. In terms of thermal performance the windows and doors in external walls are the weak point. They are, however, essential to the performance of a building. We need windows and we need doors. Chapter 16 explores the challenge of trying to achieve a consistent thermal performance of the wall. Is it possible to achieve an external wall that includes windows and doors with the same level of thermal performance, i.e. a wall and its openings as an entity with one level of performance?

Analysis and choice are collaborative endeavours. In education the examples given in the chapters that follow can be addressed individually or as small group exercises. In practice these and similar questions would be addressed collectively by the design team, working closely with the client, together with those responsible for providing materials, components, plant and equipment. The questions may also result in the need to explore appropriate digital technologies to assist with the analysis and inform the choices. We need to be thinking about the whole life cycle of the building and its components.

The case study building at a future date

If we extend our scenario to the future, we can see that the case study house will be subject to many changes before the end of its serviceable life. The building may change ownership after, say, five years, with the new owner keen to personalise the interior to their taste.

This may result in the replacement and renewal of kitchen units and bathroom fittings, which has nothing to do with technical failure. Perfectly serviceable kitchens and bathrooms will be replaced as a consequence of taste despite having a service life of 25 years or more. The replacement of doors, both internal and external, and windows may also happen because of change of building owner and/or changing tastes. Other reasons for replacement of the external doors and windows relates to weathering. For example, external timber doors may warp and become misaligned with their frame, making them difficult to open and close. The sealed units may fail in the double and triple glazed units, and over a longer time period the window frames will deteriorate and may need remedial work and or replacement. The foundations should remain sound unless there is unexpected ground movement. The 'cavity' wall should also remain sound for the life of the building, however, it may be necessary to repoint mortar joints and replace individual bricks if they happen to weather poorly, for example due to inappropriate detailing, choice of bricks and the exposure of the building to wind, rain and frost.

Foundations – Do We Need to Dig a Hole? Chapter 14 AT A GLANCE

What? The basic function of the foundations is to maintain the integrity of the structure and fabric of the construction above ground. The foundation must take account of the pattern and size of the loads from the superstructure and the stiffness of construction above ground, as this determines its ability to respond to differential settlement. Foundations will have to have sufficient area, have loads placed centrally, be stiff enough to transfer loads evenly to the ground, and be at sufficient depth to limit settlement and avoid collapse.

Why? Relying on tried and tested foundation designs usually requires the use of considerable amounts of concrete and steel for reinforcement. To reduce the environmental impact of foundation construction we could question why it is necessary to excavate a hole and fill it with concrete and steel. Are there other options? Are there other materials that could be used? What do we do with the foundations once the building has completed its useful life and has been dismantled and recycled? Can the foundations be reused for a new structure without disturbing them, or if not, can the materials in the foundations be recovered and recycled to form new products?

When? The soil needs to be investigated to determine not only the structural properties of the soil but also its moisture stability, any contamination and any suspected features that may affect the stability of the foundations. This will be done before the foundation design can commence. Once completed, the site will be prepared, and excavations will be made for the foundations to be placed.

How? Usually it will be necessary to excavate the ground to a suitable level to place the foundation materials. This requires machinery and energy, resulting in spoil from the excavation that must be used on the site or moved off the site. A simple strip of plain unreinforced concrete will be sufficient for most houses. In conditions where the construction of the simple strip carries higher risks and costs due to the required site operations, trench fill, piles, or a raft 'floated' on the surface could be considered. Steel reinforcement may also be required to provide additional strength to the concrete strip foundations, depending on the loads imposed on the foundations and the soil conditions.

14 Foundations: Do We Need to Dig a Hole?

It is the function of the foundations to maintain the integrity of the structure and fabric of the construction above ground. This chapter considers the foundations to comprise all works up to the damp-proof course (DPC) for the wall, but not including any construction associated with the ground floor that is taking loads from the activities on that floor. The foundations, therefore, have no environmental functions. Foundations are entirely structural, although they may have some minor appearance considerations in situations where the foundation wall shows above the finished ground level up to the level of the DPC. Analysis of the life cycle is necessary as maintenance targets should be zero, and the foundation will affect the possibilities in future redevelopment of the site when the building is modified or disassembled, and the site repurposed. With a few exceptions, foundations to small scale domestic properties are usually composed of concrete and reinforced with steel. The aim of this chapter is to raise questions about how we may explore other options for low rise developments, through suggestions, analysis and choice. How, for example, can the circular economy principles be applied when designing foundations?

14.1 Structural integrity of foundations

Given that the primary structural role is to maintain the integrity of the building, two things need to be known about the structure and fabric above ground:

(1) The distribution, magnitude and direction/point of application of the loads
(2) The stiffness of the construction above ground

For the house, all the loads come from walls, arriving at the foundations as uniformly distributed strip loads. Each wall will have a different load per metre run depending on which walls take the roof and floors. Wall loading for housing also depends on the self-weight of the wall, which can be a significant part of the load particularly if constructed in masonry and with few or small openings. While the loads from external and party walls vary, the difference is unlikely to be great and therefore it is usual to specify all foundations to these walls based on the greatest loading. This will theoretically lead to different settlements under each wall, but in practice, on reasonably good bearing soil, these variations in loading do not lead to significantly different settlements that would cause distress to the

Barry's Construction Technology: Analysis and Choice, Third Edition. Stephen Emmitt and Tony Bryan.

superstructure. However, the loads on internal partitions will be considerably less and the wall will be thinner, and so smaller foundations would be chosen.

House construction will produce vertical loads at the foundations and, although theoretically most of the imposed load will be from the internal skin, the load will be taken to be acting through the centre line of the whole wall. The stiffness of the building, its structure and fabric, is a measure of how much differential settlement it can take before it suffers any excessive distortion or cracking that would lead to a failure of performance. Initially, distortion or cracking may only affect appearance, but as they become more pronounced, weathering (serviceability) and structural stability may be threatened. All buildings distort and crack to some extent and much detailing is designed to accommodate these movements. It is this detailing and the choice of materials that determine how much movement can be accommodated without affecting performance. In house construction a major consideration is the stiffness of the blockwork and/or brickwork. As the stiffness of the structures increases, critical strains (the movement that develops ultimate stress and cracking) will be reached with smaller differential settlements. The use of cement mortars for strength makes the wall relatively stiff and less able to take differential settlement compared to walls constructed using the weaker and less rigid lime mortars. The nature of the connections of the floors and roof to the wall being pinned allows some rotation but they are susceptible to lateral movement, which may involve a reduction of the bearing.

Soil as part of the structure

The soil under the foundation will be stressed and is therefore part of the structural system. The foundation must act to transfer the load in such a way that neither the foundation nor the soil is overstressed. The analysis of the behaviour of the soil under load is given in Part 1. It identifies that the compressibility of the soil below the foundation to a depth of around 1.5 times the width of the foundation will determine the settlement, while the soil to the side of the foundation will be involved in a collapse failure dependent on the soil's shear strength. All this is dependent on the absence of instabilities in the mass of soil outside this area of stressed soil caused by volume change from moisture movements or soil collapse from slope instabilities or voids caused by natural or mining activity.

This analysis identifies what needs to be known about the soil and the extent of investigation that is required to establish good information on which to base the design of the foundation. The basic soil properties of the shear strength to determine the ultimate bearing capacity and the compressibility (including time-dependent behaviour) must be known. In addition to these properties that are related to the capacity of the soil to resist the loads it is important to know about the soil's volume change with changes in moisture content. It is characteristic of clays (and the more cohesive silts) that changes in moisture content are accompanied by changes in volume. Removing water from these soils below foundations leads to shrinkage that would cause settlement additional to that expected from the application of loads. Increased water content in these soils causes swelling. If additional water becomes available, the suction forces in the very fine structure of cohesive soils are great enough to overcome the weight of the building and the ground will heave, lifting the building up. Once it is established that the soil on the site is subject to these volume changes, it is necessary to ask to what depth these moisture changes can be expected.

There are two major causes of moisture change: seasonal changes and the vegetation (mainly trees) around the building. In the yearly cycle between summer and winter the soil near the surface will dry out (desiccate) from the sun and take up moisture again with the increased incidence of rain. In the UK, these seasonal changes are variable, but over the life of the building some dry summers and wet winters will be experienced. Past records have shown that these are unlikely to affect the soil below about 900 mm so any foundation below 1000 mm should not experience this seasonal volume change in the life of the building. Vegetation, however, can extend its influence below this level. The effect is very dependent on the species of the tree, its thirst for water and its mature height, and how far it is from the building. Mature trees will draw water from deeper and wider sources in dry summers, leading to shrinkage and settlements. If mature trees are removed, they will cease to take water and the soil will return to a higher water content, causing swelling and heave. The depth of soil affected must be assessed for each site and incidence of trees but can be as deep as 3500 mm or more, much greater than seasonal influences.

There is one other volume change associated with the presence of moisture, but this is not from change in moisture content but in the freezing of any moisture in the soil. Soils susceptible to movement under these conditions include silts, chalk, fine silty sands and some lean clay. This can cause heave if freezing takes place below the foundation. For housing this is a seasonal change associated with very cold winters with prolonged sub-zero temperatures. Again, the winters in the UK are very variable in this respect, but past records have shown that in areas adjacent to occupied buildings depths of up to 450 mm have been affected, so foundations below 500 mm should not be affected. For specialist installations such as cold stores, greater depths may have to be considered.

Site investigation

It is now possible to identify what needs to be known about the soil to a depth that includes the zone of stress and moisture changes if the soil is expected to be clay. It is also necessary to be clear how these can vary under the building, including the possible shear zone around the building. If the position of the building is not known, the whole site must be considered in the investigation. Beyond this, other information is necessary to consider the possibilities of subsidence or other ground collapse, which may be a potential failure to be allowed for in the design.

In establishing a site investigation it is also necessary to gather information important to the production process. Work in the ground carries with it risks directly related to the ground conditions. These are not only technical risks but also health and safety and financial risks, which tend to increase the cost of works below ground. Selecting foundation solutions that minimise these risks is part of the process of choice. The soil properties identified for the design will give information on the risk of collapse of exposed excavations and the need for temporary support, but additional information, such as ground (standing) water and access possibilities, needs to be established.

Information for environmental analysis must also be collected at this time. This could be concerned with habitat and the local ecosystems as well as previous land use and the possibility of contamination. Care must be taken in making this distinction between green and brownfield sites. Some areas of the UK now considered wild and unspoilt might have had an industrial past from Roman and even medieval times. Previous development may indicate the existence of old foundations or other construction below ground or contamination. It may also reveal antiquities or other archaeological remains.

Not all information can be gathered in the same way. This leads to most investigations being based on some or all of three processes:

- Desk study
- Walk-over study
- Ground investigation

The desk study uses existing published information. This can include climatic data, but it is most likely to be based on geological and topographical maps, aerial photographs and in some cases ground penetrating radar. Analysis of this data will yield information to assess the wider possibilities of subsidence from both geological landforms and previous human activities that could make the site unstable. This would therefore include records of maps or old mine works that would show previous land use. The names of features or settlements are often clues to previous use. The desk study would also include surveys of the utilities, such as water, electricity, gas and sewers, which can be obtained from the appropriate authorities. Risk of surface water flooding can be obtained via the UK government online mapping service.

The desk study will indicate what may be found on the site and will inform the other two phases of on-site investigations. Along with any local knowledge, it will give some idea of what land features and vegetation would be visible from above ground and what soil type, previous development and contamination might be below ground level. What is considered contamination could come from two concerns: the adverse effect it may have on the health of the occupiers (and constructors) of the building and the possible aggressive effects it may have on the materials to be chosen for the construction. This will determine what could be considered sufficient levels of concentrations in the soil or ground water to be considered contaminated. For these reasons, contamination surveys on brownfield sites are often undertaken independently of the engineering soil survey.

The walk-over survey gathers information from above ground. This would include the topographical and vegetational features and the conditions of adjacent development. It may be possible to just mark these on an existing site plan plus a photographic record, but it may show the need for a more accurate land survey if the positions of features or the exact levels of the land were important to the proposed building, its form and position.

The scope of the ground investigation will depend on what the desk and walk-over study identified and the size and possible construction of the building. For house foundations, simple trial pits will probably be sufficient, with the soils being identified with the use of soil classifications from visual inspection. If the site provides good bearing conditions, simple strip foundations can be designed to satisfy the Building Regulations from a table in the Approved Documents, where only soil type is needed to select a foundation width. Simple visual classification will also be sufficient to determine the presence of the shrinkable clays important to specifying sufficient depth to ensure volume stability. Trial pits will give an indication of standing water and will be sufficient to take water and disturbed soil samples for chemical analyses, such as sulphates, if these are suspected. Where visual identification of soil type is not sufficient, *in situ* tests and the removal of undisturbed samples for laboratory tests, possibly from boreholes, may be required.

It is important to consider a fourth phase that takes place during construction where the conditions anticipated by the site investigation and used for the design need to be continually compared to the information revealed by excavations for the building. This will also lead to inspections prior to significant events, such as pouring concrete for the foundations.

14.2 Foundation design

Whether the foundation designs are chosen from tables or are the subject of a structural design, four conditions must be satisfied:

(1) The load must be distributed over a sufficient area to limit settlement and avoid collapse.
(2) The centre of gravity of the load must be over the centre of the area of the foundation.
(3) The foundation must be stiff enough to press evenly on the ground over the whole area of the foundation.
(4) The foundation must be founded at a depth below ground to be resting on a sufficiently stable soil mass.

For the house under consideration in the case study the loads are relatively low, and many sites in the UK have soil with a good bearing capacity and low compressibility, although it is more than likely to be shrinkable clay in England, particularly in the south and east. If there are no major previous developments or geological or typographical difficulties, a simple strip footing may be the economic choice.

The strip foundation

The basic form of the strip foundations is shown in Figure 14.1, with a cavity wall construction. The cavity wall sits on a DPC to stop rising damp. The DPC must be at least 150 mm above the ground to avoid soaking of the wall from splashing back from the ground. The

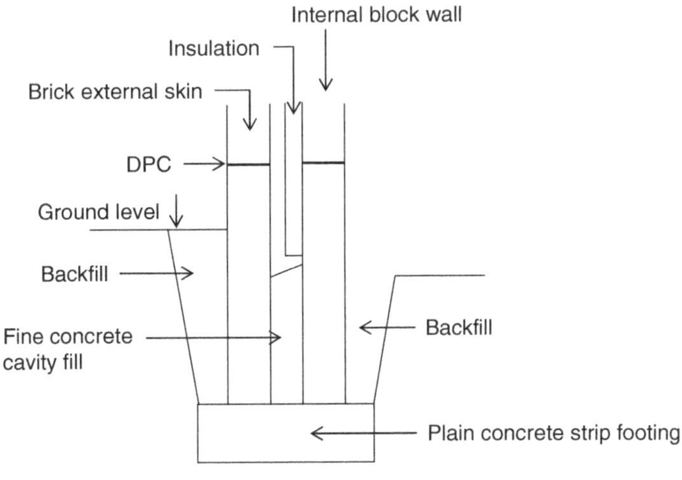

Figure 14.1 Plain concrete strip foundation.

first part of the wall below the DPC must remain as a cavity wall to ensure that the cavity remains clear and can drain below the DPC. The external skin must remain brickwork for at least two or three courses below ground to maintain the appearance, but this assumes that the brick will remain durable in this zone where saturation and freezing are more likely. Below this, the external skin can be constructed in blockwork. Nearly all blocks are suitable for construction below ground, so a change from the insulating block used in the inner skin above the DPC can be made if this does not create a thermal bridge at the edge of the slab.

From 150 mm below the lowest DPC the cavity must be filled, as the wall will be subject to pressure from the soil tending to collapse the two skins together. It is possible to use a single trench block along the total width of the wall to avoid having to build the two skins and then fill the cavity. The bottom of the wall then needs to be stiffened and probably made wider to transfer its load evenly to the soil. If the width does not need to be too much wider than the wall, a strip of plain (unreinforced) concrete can be used. This is the case for many sites in the UK, where foundation widths between 300 and 600 mm will be sufficient. This concrete will not have to have great strength but will need to be durable and therefore the specification needs to include the minimum cement content (and type if sulphate resisting is required), as this will be governed by durability considerations.

Figure 14.2 shows how the design principles given above are observed. First the load and type of soil will determine a width to provide sufficient area. The wall must then be detailed centrally on the strip. To provide a stiff base to the wall in its length the footing must be at least 150 mm thick. However, to ensure the load is transferred across the whole width of the strip the thickness must be equal to the projection. This is related to the choice of the material as plain concrete. The strip will transfer the load to an area under the footing if the material remains in compression inside the 45° punching shear line.

This requirement for the thickness to equal the projection will give thicker foundations as the width increases. As the thickness increases it may be worth considering the possible change to the material whose structural behaviour is not dictated by the 45° punching shear failure in plain concrete. This can be achieved by introducing steel reinforcement into the concrete to provide a 'wide' strip foundation, as illustrated in Figure 14.3. The foundation can now sustain a bending action, ensuring the load is spread across the full width

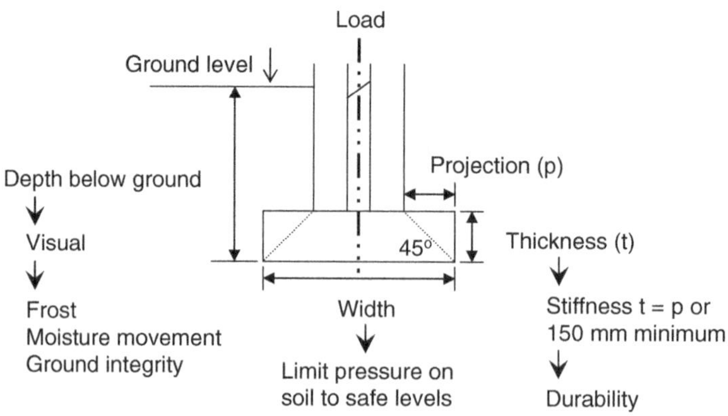

Figure 14.2 Design considerations for plain concrete strip foundation.

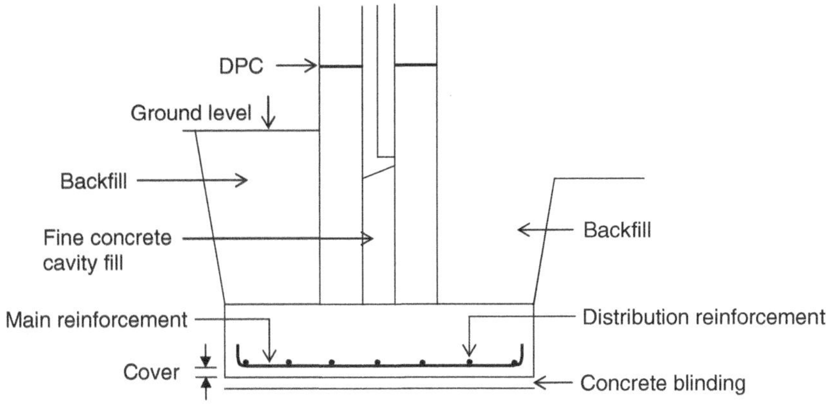

Figure 14.3 Reinforced concrete 'wide' strip foundation.

of the strip. The introduction of reinforcement requires the specification of cover. This is the thickness of concrete required to gain a bond between concrete and steel to develop a composite action in bending and to protect the steel from corrosion (above ground this may also be determined by fire requirements). For foundations this is determined by corrosion protection and will be 40–50 mm, depending on conditions. To ensure this cover is achieved in the production process a concrete blinding is required and the reinforcement will be supported on spacers standing on the blinding. In this way the reinforcement will not be trodden into the ground during the concreting operations.

Returning to Figure 14.2, the last parameter is the founding depth below ground to ensure the foundation is resting on a sufficiently stable soil mass. Given that the top of the strip must be at least two or three brick courses below ground, it is unlikely that the depth of the foundation (to the underside of the strip) will be less than 300 mm, so it should be below the level of the topsoil (normally around 100–150 mm in the UK). However, this is not below the frost line, requiring depths of 500 mm, or the influence of seasonal moisture variations on clay at 1000 mm, so these may determine the depth on soils susceptible to movement under these conditions. Deeper foundation may have to be considered if there is vegetation on shrinkable clays or on any site where geological, old industrial or topological features must be considered.

The simplest topological feature is a sloping site. Even if it is not sufficient to make the site unstable, it will affect the foundation depth. The specification of depth below ground will have to be for the lowest part of the site, so, if the whole strip is laid to the same level, the depth of the foundation will increase at the rate that the level of the ground rises. As the depth increases the production risks increase. There is a greater need for working space and temporary earthwork support. This will need larger machines and require the removal of more soil. All of these increase the health, safety and environmental risks, hence both cost and time will increase. These concerns need to be considered beyond around 1200 mm below ground level. To limit the depth on a sloping site the foundations can be stepped, as shown in Figure 14.4, where the dimensional requirements are for a plain concrete strip.

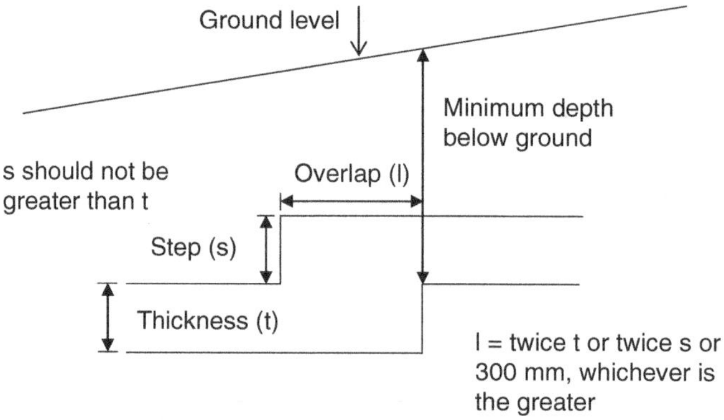

Figure 14.4 Steps in plain concrete strip foundation.

Foundations for poor surface conditions

Apart from sloping sites, other conditions lead to the need to consider deeper founding depths; these include:

- Potential volume change in shrinkable clays
- Variability of the soil across the site
- Surface soils with poor bearing capacity

The production costs and risks will have a great influence on devising solutions to provide deep foundations. Most systems for deep foundations have been developed to be installed from ground level to limit the health and safety risks, construction time and cost. However, the answer may not be to provide deep foundations at all but to support the building at the surface in such a way as to limit the pressure on the ground and stiffen the foundations to limit the effect of differential settlement on the structure and fabric of the building.

Solutions available fall into three categories:

(1) Foundations to depths around 4500 mm
(2) Foundations to greater depths
(3) Surface 'floated' rafts

Deep foundations

Solutions for depths to 4500 mm have been mainly developed for sites with problems associated with shrinkable clay, but they can be considered for any sites with good bearing ground within this distance from the surface. Trench foundations can be considered for soils that can be excavated without the need for earthwork support within the first few hours of excavation, for example clay soils. Regardless of how deep the founding level, the concrete is filled up to a level that allows the safe and economic laying to the wall up to the DPC, as shown in Figure 14.5. In this type of foundation the production conditions are the key. The trench must be able to be excavated accurately (line, level plumb and width) without the risk of collapse. This must be achieved in a reasonable time to excavate sufficient length of

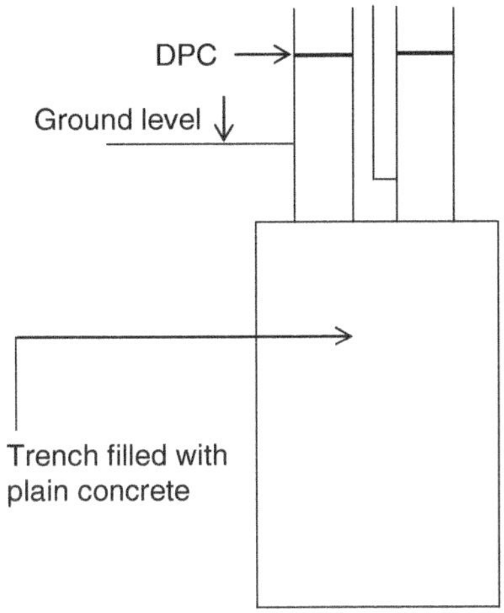

Figure 14.5 Plain concrete 'trench fill' strip foundation.

foundation, confirm the founding depth and pour the concrete to return the ground to a safe condition. Generally this cycle of operations will have to be completed within one day. These conditions can be fulfilled in most clay soils, but the time factor may limit the depth to which this form of foundation can be undertaken, as the longer the trench is left open, the greater the risk of collapse as the moisture content of the clay changes. If the trench can be cut cleanly and accurately the quantity of concrete can be controlled, limiting the financial as well as the health and safety risks.

If there is then a risk of heave (expansion) in the mass of soil contained under the building between the foundations, a compressible material should be introduced to the face of the foundation, and the ground floor possibly constructed with a void as a suspended floor.

An alternative to trench fill is to provide piles to take the loads down to the safe bearing soil with a beam across the top of the piles to support the wall. Up to depths of around 4500 mm this can be achieved with what are known as short-bored piles, as illustrated in Figure 14.6. Here short and bored are determined by the production requirements. Piling is achieved either by removing the soil and then replacing it with concrete (replacement piling) or by driving the pile to displace the soil (displacement piling), and these can be achieved in timber, steel or concrete. In replacement piling the method of removing the soil is by boring or drilling a shaft into the ground. This process will produce a circular shaft, the diameter and length of which will determine the size of plant (piling rigs) that will be required. By limiting the length to around 4500 mm and the diameter to around 350 mm, piles can be achieved with relatively small rigs. If greater depths and/or larger diameters are required, full piling rigs are required.

Even with a circular shaft with a limited diameter, some soils may collapse into the hole before there is a chance to fill the shaft with concrete. In this case the shaft must be created with a tubular lining normally recovered as the pile is concreted. Displacement pile systems with precast piles and beams have also been developed where this risk of collapse is considerably reduced so long as founding depth is well established and verticality during driving can be achieved. Both conditions can usually be found on clay sites.

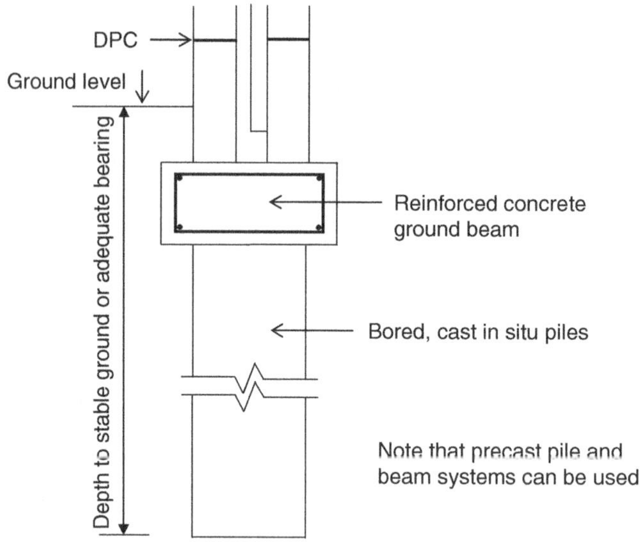

Figure 14.6 Pile and beam foundation.

The beam must span between piles to support the wall. Piles are placed at relatively close centres of around 3 or 4 metres, limiting the need for a deep beam. In addition, piles are placed at corners of the building and at the ends of door and wide window openings. Production operations to create the beam can be achieved in a trench near to ground level. Here the beam must be wide enough to carry the wall, so the characteristic shape of the beam is wider than it is deep, not the usual proportions of a beam. If these foundations are in shrinkable clay with the beam near the surface, shrinkage and heave could take place under the beam and around the top of the pile. The beam must be cast on a compressible layer, and in some cases there may need to be a slip membrane around the top section of the pile.

Where founding depth is greater than 4500 mm, the need for longer piles necessitates the use of full-size piling rigs, but the principles are the same. With the relatively light loads from houses, the need for deep piling is unusual. There are several options of both replacement and displacement systems to create the piles.

The idea of the short-bored pile can also be achieved with a pier and beam foundation. This involves the excavation of pits for pad foundations where masonry piers are constructed to take the beams that support the wall. This requires men to work at the founding level with the necessity to provide working space and health and safety measures.

Raft foundations

The alternative to providing deep foundations is to investigate the use of surface 'floating' rafts. These combine the ground floor and the foundations into one continuous structural unit. All the loads combine and are spread out under the raft so that the entire footprint of the house is the foundation bearing area. This will create very low pressures on the soil under the raft, although it will stress the soil to a greater depth, albeit that they will be low stress values. This gives a safe solution on soils with limited bearing capacity and has the potential to accommodate differential settlement without transferring the strains into the superstructure. If differential settlement occurs, the raft can be designed to tilt with no strains induced in the superstructure (this can be successfully used in areas subject to mining subsidence). This can only be achieved if the pattern of loading is relatively even to give the centre of gravity of the loads at the centre of the area of the raft and if the construction can produce a raft stiff enough to act as one foundation.

The stiffness or rigidity of the raft is the key on sites where there is variability in the soil strength and/or compressibility under the building, giving the possibility of differential settlement. As the foundation to a house is a limited area with a relatively high perimeter/floor area ratio it should be possible to achieve this rigidity by stiffening the edge where the main loads will be taken from the external walls. Thickening may be necessary under internal loadbearing walls. As the ratio of perimeter to floor area decreases it may become economic to stiffen the internal area of the slab with beams rather than just increase the thickness of the slab.

For housing, the slab is likely to be between 150 and 300 mm thick with an edge stiffening detail dependent on the soil conditions. For soil of medium to low compressibility that is consistent under the whole raft the edge detail may only need to be a reinforced extension beyond the external face of the wall. These details are shown in Figure 14.7. This extension is particularly significant when the raft is constructed close to the surface. Relatively small extensions can give some protection from frost heave but may have to be greater if the raft is on clay, with the risk of shrinkage and loss of support under the edge of the raft. On granular soils a small downstand provides the best protection, as erosion may progressively undermine the edge of the raft over time.

On soils of high compressibility but particularly where the soil is variable over the site effective support may be reduced over some sections of the foundations, and there is a need to provide additional bending capacity as well as protection. Here the downstand beam detail shown in Figure 14.8 will be required.

All rafts are reinforced with mats for the slab that is usually provided with pre-formed mesh reinforcement. For light edge downstand details pre-bent mesh cages may be considered, but for edge beam raft design the beam cage will need to be specified with bars and

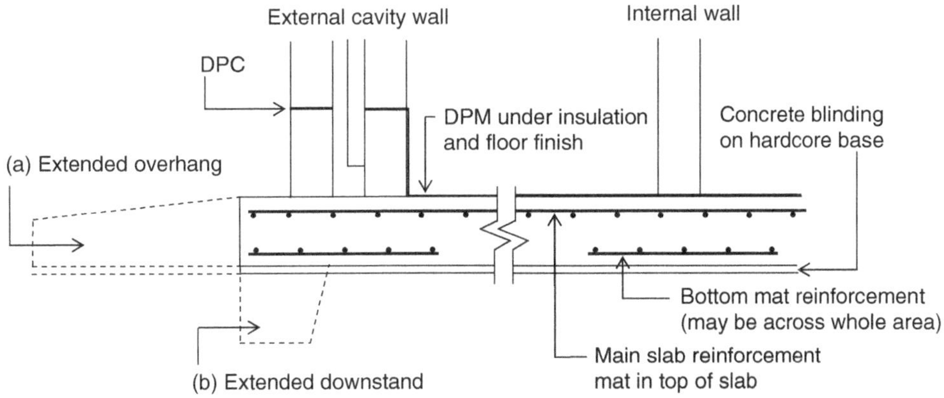

Figure 14.7 Plane raft foundation.

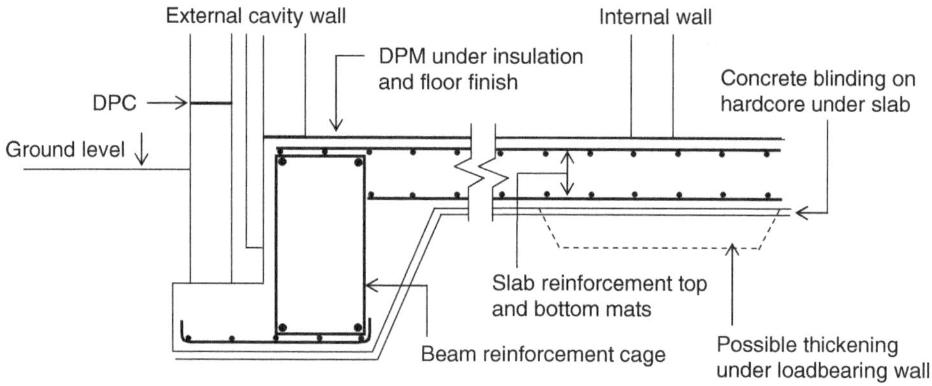

Figure 14.8 Edge beam raft foundation.

links that may be fabricated on the site. As with the 'wide' strip foundation the reinforcement will need to have sufficient cover, and to ensure this is maintained in the production process there is a need for 50 mm of concrete blinding and the use of spacers and chairs to maintain the position of the top mat in the slab. The edge of the raft will have to be formed using formwork with some support falsework to hold the form in position during concreting. This will determine the accuracy of the line and level of the concrete that will determine the ability of the bricklayer to achieve the tolerance of the external brickwork.

14.3 Economics of foundations

The evaluation of build costs for foundations needs to be done in relation to the elements of risk and the variety of site conditions. These can be a major determinant of the choice of foundation method. Depth is not the only factor to consider. If the production process for deep foundations could be carried out from the surface with little risk, they might be less expensive than a raft with formwork, with the steel reinforcement and concrete making up the major cost. For the strip foundation options, when to use wide strip or trench fill rather than the simple strip will again be dependent on the site conditions and the risks involved. To this can be added the possibility of using ground improvement techniques to increase the bearing capacity of the soil particularly in non-cohesive soils. This may make simpler, less expensive foundation viable, offsetting the cost of the ground improvement process. Techniques such as dynamic consolidation, the installation of stone columns and the introduction of cementatious materials into the ground can improve ultimate bearing capacity to economic depths for strip foundations to be used for housing.

As noted in this book, the initial build cost is not the only consideration. Unlike the main building structure, which is exposed to weathering, there should be little need to maintain the foundations because they are protected from the environment, so the maintenance costs should be zero unless some of the foundation is visible above the finished ground level. In some cases there may be local failure of foundations due to unexpected movement within the ground on which the building is founded, which will require repair to ensure stability of the structure is maintained. The third cost to consider is the future recovery of foundations and materials once the building has been disassembled at the end of its serviceable life. This requires life cycle costing (LCC) to establish the whole life cost of the foundations and needs to be analysed against the plan for disassembly and materials recovery.

14.4 Exploring some scenarios

By describing and illustrating typical construction solutions for foundations for a small house in the UK it is possible to see some common characteristics. The foundations all rely on concrete that is reinforced using steel reinforcement bars (commonly known as rebar). They are also reliant on excavation of ground to provide a stable base onto which the foundation is constructed. This uses resources and energy. Hence it may be instructive to ask the question, is it possible to use foundation designs that may be more environmentally sustainable compared to typical and common solutions illustrated above? The question raised and the potential solutions may challenge existing knowledge and good practice guidelines. The designs would require input from structural and civil engineers prior to seeking approval from building control. This typically takes longer and requires additional specialist knowledge compared to adopting a typical solution. There is also an increased element of risk, which often results in designers and engineers over-specifying materials 'just in case' to avoid the prospect of failure. In this situation the tendency to over-specify may result in additional, not fewer materials being used, which may defeat the purpose of the exercise.

A small number of simple questions are raised in this section as a means of stimulating our thoughts.

Do we need to dig a hole?

Depending on the site conditions, the answer may be no. To ensure that foundations are more environmentally sustainable there are solutions being tried that do not rely on digging a hole. Some sites may be suitable for positioning the foundation on the ground without the need for excavation of the ground. This would usually require a concrete raft foundation that sits on the ground. This would need to be protected against water and frost damage. For lightweight structures, for example a timber or a steel framed house, it may be possible to use short-bored concrete foundations with a steel or timber support placed onto it so that the building sits slightly above the ground. This results in minimal disturbance to the ground. For sites that may be prone to flooding because they sit within a known flood plain and/or may be at risk from flash flooding during very heavy rainfall, this system could be used to raise the building off the ground to a level that should not flood.

If steel screw piles are used it is easier to remove them once the building is no longer required. This type of solution would suit offsite construction and may result in a quick build time on the site compared to using traditional concrete foundations and loadbearing construction. Depending on the condition of the steel piles when they are recovered, they could be reused on the site by repositioning them or they could be transported to another site for reuse. We can see here that in thinking about the foundations also has implication for the weight and type of building that is to bear on the foundation as well as its production methods. This also prompts decision makers to think about the longer-term sustainability of the building, its estimated design life, and what is to be done at the end of the building's useful life. It also raises associated questions relating to performance in extreme weather, in this case flooding.

When building in areas that are prone to flooding, raising the building off the ground, and above the highest predicted flood level, is one solution. Another is to embrace the water and use floating foundations as has been done successfully in, for example, the Netherlands. In this case the building will sit on a floating raft that is secured to a pile, and which allows the building to rise and fall with changes to the water level. This raises questions about how services are to be connected to the building to allow for the vertical movement and how the surrounding infrastructure of roads and/or waterways will allow access and egress to the property if the immediate surroundings are flooded.

Do we need to use concrete and steel?

Concrete is widely used throughout the world for foundations, but it is not a very environmentally friendly material. The production of concrete produces carbon dioxide, uses energy and requires a significant amount of water. Although manufacturers are taking steps to reduce the amount of energy used and the amount of carbon dioxide produced during the production of concrete, concrete is still responsible for a significant amount of greenhouse gas emissions. Efforts to reduce the carbon footprint of concrete have resulted in the use of the term 'sustainable' or 'green' concrete to cover the manufacture of concrete that incorporates waste material or recycled content to reduce the amount of carbon dioxide produced during manufacture. Whether these new concrete mixes are 'sustainable' or 'green' is very much a personal interpretation of the marketing surrounding all materials and the ongoing efforts to reduce their negative impact on the environment.

Steel reinforcement (rebar) is common to concrete structures. Like concrete, steel production uses a high amount of energy in its production and there is an initial high environmental cost associated with steel production. This can be reduced through recycling and the reuse of scrap steel recovered from the disassembly and demolition of buildings. Steel is also prone to deterioration, especially if the depth of cover provided by the concrete is not as specified. Other materials may be suitable for foundations, and research is ongoing into the use of fibre-reinforced polymers. This relies on the use of micro or macro synthetic fibres in the concrete mix to replace the need for steel reinforcement. Fibres can also be made of plastic, carbon, glass or steel. Engineered bamboo is also being trialled and researched as an alternative to traditional steel rebar. Some of these alternatives to steel reinforcement may present a practical and sustainable option for our case study house. We do, however, have questions relating to the future recovery of materials and the long-term structural performance of these relatively new options.

Can we use less material?

Designing foundations to use less material than would be required to comply with structural codes and standards is challenging and poses additional risk to that normally deemed to be acceptable. We do know, however, that many structural components such as floors and foundations are over-designed to allow for a safety factor. The question needs to be asked as to whether the safety factor is too high. If it was lowered, would it result in a reduction of materials that would improve the environmental impact of the construction solution?

There may also be initial cost and time savings. To address the over-design of foundations would need a more nuanced approach to foundation design and a move away from tried and tested solutions that are known not to fail.

Can we reuse or recycle foundations?

Reuse of existing foundations can help to reduce carbon emissions associated with plant and machinery, transportation and production of concrete and rebar. Removing a building and recycling and reclaiming materials at the end of its useful life will leave behind the foundations. It has been common practice to dig out the foundations and crush them to release aggregates and to reclaim steel reinforcement for recycling and reuse. The steel will need to be taken offsite for processing. In some cases, the aggregate recovered from the foundations can be used on the site, for example for hardcore, hence avoiding transportation of materials elsewhere. If the foundations are to remain in place, and they are still structurally sound, they can be reused. This requires very careful design of the new structure to fit the exiting footprint of the foundations, and in many cases it may require some additional foundations to suit the new building design. Compromises in the design may be necessary to achieve a low impact building on an existing footprint. This brings us back to analysis and choice and the need to make trade-offs to achieve our desired goal.

Reflective exercises

A conscious decision has been taken to not include any drawings, sections and details, to allow readers to develop their own conceptual details in relation to foundation design. Continuing with our case study building, how would the details look if we embraced some of the thinking outlined in Section 14.4?

Exercise 1. Try drawing a section of a foundation design from first principles without relying on existing details to be found in the construction technology books. Think creatively, do not be bounded by convention.

Exercise 2. If Exercise 1 proves to be too challenging, try taking an existing detail, for example one of those provided in this chapter or in *Barry's Introduction to Construction of Buildings* as your starting point. Then analyse the chosen detail and explore what could be changed to create a foundation design with better environmental credentials. How would you justify your decisions?

Walls – The Paradox of the Cavity Wall: Chapter 15 AT A GLANCE

What? The loadbearing masonry cavity wall, chosen primarily for its weatherproofing action, is the most likely option for the external walls for the house in the case study. The cavity that provides the weatherproofing is formed between an outer skin that determines the appearance and an inner skin that normally takes the loads. In loadbearing masonry construction, the cavity and the choice of lightweight blocks for the inner skin provide some thermal insulation, but this must be augmented with a layer of insulation to achieve the required U value. The insultation is inserted into the cavity, which reduces the depth and effectiveness of the air cavity, and in many cases fills the cavity completely.

Why? The cavity wall was developed to limit the risk of damp penetration through solid masonry walls to the building interior. This could be a problem with solid walls in exposed conditions, particularly via the joints, and aggravated by poorly maintained pointing and/or brickwork. The principle was to provide an outer masonry skin that limited, but did not have to eliminate moisture penetration under all conditions. The cavity, a clear air space, provided the guarantee of damp-proofing if the cavity was not breached. The inner skin then remained dry and could be used to support the interior wall finish without fear of damp within the building.

When? The cavity wall will be constructed using lifts of brickwork and blockwork, building up from the foundation. The actual sequence will depend on building traditions and preferences of the bricklayer; however, it is common to build the internal blockwork leaf and fix rigid insulation and wall ties, then the brick (or stone) external leaf is built, hence completing the wall. This is done in 'lifts' with the external walls built to the height of the first floor. Once the floor timbers have been added they provide sufficient rigidity to allow the next lift of blockwork and brickwork for the upper storey of the house.

How? In the UK masonry cavity wall construction has evolved from a relatively simple form of construction to one that is complex, and which requires additional skill in its construction compared to earlier versions if the performance criteria are to be met. There is no longer a 'cavity' and the construction may be better described as a sandwich construction. The failure of the external wall to meet thermal insulation performance targets is usually down to errors in workmanship. Given the complexity of the wall construction this is not uncommon. Thus, the fundamental question to ask is why do we need to use a masonry cavity wall? Although this form of construction is very familiar, it does not mean that we should keep using it if it has outgrown its original function.

15 Walls: The Paradox of the Cavity Wall

In the UK, the cavity wall has become a common form of construction in domestic scale construction. This form of external wall is common in loadbearing masonry construction and is also used in many framed constructions, such as timber and lightweight steel systems. Given the widespread use of cavity walls, it is likely that this form of construction would be chosen for our case study. In this chapter the focus is on masonry cavity walls. Openings in the cavity wall for doors and windows are addressed in Chapter 16.

15.1 Walls by function

Walls play a major part in the construction of a house, providing a wide range of functions dominated by the structural and environmental roles that they are asked to perform. Walls are given different names depending on the role they play in the overall design based on the broad set of functions they perform. We use the terms loadbearing and non-loadbearing walls, based on the direct analysis of the load paths from dead and imposed loads. This is a simplistic and limited definition of being part of the structure, as overall stability associated with wind loading, and component instability often relies on walls that may not be carrying the direct loads from the construction and the occupation of the building. Many walls that are called non-loadbearing are part of the structural system and therefore are subject to stress under certain conditions.

Environmentally we talk about external walls, separating (party) walls and partitions (internal walls), each having an associated set of functions that will determine the choice of construction. Given the exposure conditions in this case study, external walls are dominated by the needs of weathering and insulation to resist the passage of heat. Openings for windows and doors are a major influence on the design in so far as they fulfil the functions of access, light, views, security and ventilation required of the wall. They also have a significant impact on the appearance of the building. However, the creation of the opening is complicated by the need to maintain the function of the wall as an entity. The components associated with windows and doors will have limited performance in strength and insulation, functions that were all carefully incorporated into the basic wall construction.

Separating walls are the walls used between dwellings in terraced and semi-detached house designs. They will have no openings and no weathering requirements but will need to have functions associated with security and privacy in the need for fire and sound resistance. The other internal walls within the dwelling are known as partitions. There is a need

Barry's Construction Technology: Analysis and Choice, Third Edition. Stephen Emmitt and Tony Bryan.

for openings, and the dividing function will again be dominated by privacy and security. Acting as a visual barrier will not be sufficient as the wall will need some fire and sound resistance and must be robust and capable of supporting finishes, cupboards and shelves and integrating services.

15.2 General forms

We also refer to the general forms of walls in several ways. In housing, one distinction commonly made associated with weatherproofing is reference to the external cavity wall or the solid wall. The external cavity wall has gained almost universal acceptance, although the need for increased insulation and other energy-saving ideas and perhaps a wider acceptance of rendering as an external finish has revived interest in the solid wall, but it has yet to become widespread in practice in England.

For either cavity or solid walls, three generic forms can be employed in the construction of walls:

- Masonry
- Panel (framed or solid)
- Monolithic

The basic distinction is in the size of components and their spatial arrangements, which determine the types of materials, jointing and production process involved. All walls need to provide a continuous form. If they are loadbearing walls, they need to be capable of distributing loads evenly throughout the structure. They also need to provide continuity for environmental control plus the capacity to have openings formed within the construction.

Masonry is characterised by heavy materials such as stone, fired clay or concrete, that lead to components such as bricks and blocks. These are normally relatively small units compared to the dimensions of the wall. They are regular in shape to form the bonding patterns that give stability and continuity. The choice of materials and the size and regularity of the components vary greatly in their properties, appearance and cost. Given their weight, locally available materials have in the past been chosen on economic grounds based on the cost of transport, and this has led to geographically recognisable styles (vernacular architecture). With modern manufacturing methods and the relatively low cost of transport, fired-clay bricks and concrete blocks are the dominant choice in most areas of the UK.

An external cavity wall constructed with a concrete block inner skin and a brick external skin is shown in Figure 15.1. The bricks and blocks have a modular size that includes the 10 mm mortar joint and that allows the bonding patterns (stretcher bond shown) and the use of bricks and blocks in the same wall. This basic relationship is shown in Figure 15.2. For the brick, this modularity is extended to its third dimension. The modular width is 112.5 mm, half the modular length, which allows thicker walls to be built by laying the brick with the header to the face of the wall. This full modularity of the brick is shown in Figure 15.2. Blocks are not modular in the third dimension, being available from 75 to 200 mm thick. Local availability may influence choice in conservation areas, and locally sourced materials with low transport needs can reduce energy use (rather than cost) and can contribute to sustainable development by supporting local businesses.

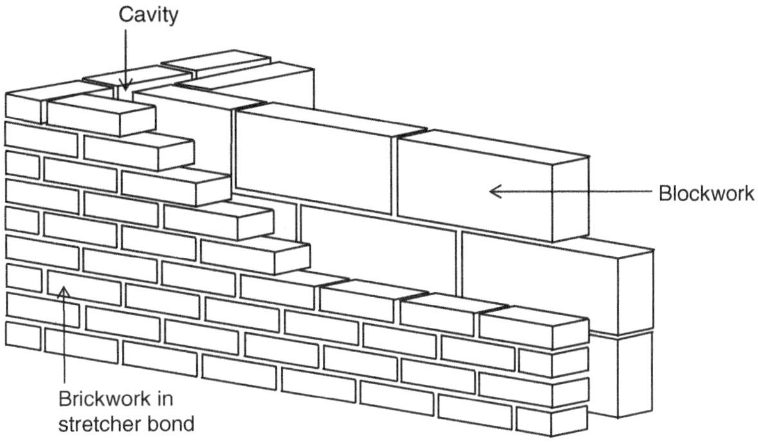

Figure 15.1 Masonry walls in brick and block.

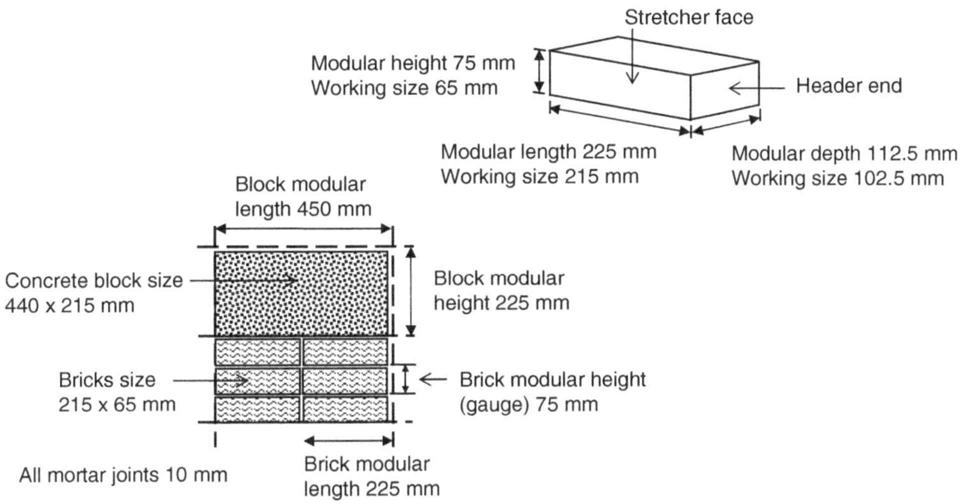

Figure 15.2 Masonry brick and block modularity.

15.3 Walls in the case study

For the house in the case study the initial suggestions to be considered will be a cavity wall with masonry external and internal skins. Identifying which of the walls are loadbearing or more generally acting as part of the structural system in a building can only be achieved for a particular design. This case study is not that specific, but it is reasonable to assume that the external cavity walls will be loadbearing and at least some of the internal walls (not considered in this case study) will also have to carry loads from upper floors. These and possibly other walls will then have to take the racking forces from wind loading. Our focus here is entirely on the external masonry cavity wall.

15.4 Masonry cavity external walls

The cavity wall was developed to limit the risk of damp penetration, which could be a problem with solid walls in exposed conditions, particularly via the joints, aggravated by poorly maintained pointing and with the introduction of strong cement mortars. The principle was to provide an outer skin that limited but did not have to eliminate moisture penetration under all conditions. The cavity, a clear air space, provided the guarantee of damp-proofing if the cavity was not breached. The inner skin then remained dry and could be used to support the finish without fear of damp within the building.

The simple principle of the cavity has proved very robust, so long as attention is paid to certain details in the wall and at openings (see discussion later). Several materials can be used for the outer and inner skin, but for the case study the initial suggestion is to choose brick for the outer skin and concrete block for the inner skin. From experience the thickness of the internal and external skin can be expected to be around 100 mm. Given the scale of house construction, this would seem to be a reasonable estimate, but the actual size will have to be determined by the performance requirements and the materials chosen. If the thickness deviates far from this size, the additional material will have to be justified from additional performance or economy and details will have to be reassessed. The width of the cavity is the other key part of the specification of this form of wall. The cavity only works if the width of the cavity at all points can be maintained within tolerances and kept clean.

These are both production problems, so the production process dictates a minimum size for the cavity. These are, therefore, related to the materials specified for each skin; in this case both are masonry for which a 50 mm clear cavity would be the usual proposal. It is not possible to provide an analysis to check this width until the whole wall specification is chosen. The size will also be determined by the introduction of insulation, so this also must be established before the size of the cavity can be determined. It will be necessary to return to confirm this initial suggestion of 50 mm when the wall specification is more advanced.

The external skin

The initial suggestion for the external skin will probably be based on its appearance. Clearly this not only will be determined by the style and status of the building but also will be a major factor in the building's response to its context. It may even be determined directly by the client who may have very strong views on what the house should be made of. Dominating the UK market is brick, although locally stone might have been historically used, so reconstructed stone blocks have been adapted for cavities. Renders might have been commonly used where local quality bricks were not available or in extreme exposures, such as coastal areas, and this again can be easily adapted for the cavity wall where concrete blocks can be used in the external skin as the backing for the render.

Exposure is a key factor in selecting the material for the external skin, less for damp penetration as the cavity has this function but for durability and potential maintenance. If masonry is to be used, the exposure to rain at times of low temperatures will make frost action a significant risk. Frost resistance will have to be considered not only in the brick but also in the choice of mortar. Other risks of change over time include discolouration. This change in colour may be desirable as weathered materials are often more mellow and attractive than when new. However, some discolouration of, say, pastel-coloured renders may not stand the test of time in the UK climate. Detailing to ensure limited pattern staining with good even washing from the rain needs to be considered.

The facing brick provides a good range of options, most of which can achieve frost resistance and have a wide range of colours. If local materials are preferred, the choice may be more limited. The modular size of the standard UK brick and block was illustrated in Figure 15.2. The brick is 102.5 mm thick so, if laid, stretcher bond can form an outer skin, around the 100 mm originally estimated. This 'half brick' skin will limit moisture penetration to the cavity but will not eliminate it, so detailing in the cavity will have to consider the presence of some water on the inner face of the brickwork's outer skin.

The structure and the internal skin

It is not possible to proceed to accept this specification for the external skin that was based primarily on appearance, or start to think about the internal skin, without being clear as to how the loads are distributed across the wall. This will be determined by the detailing from the connections of the floors and roof. Figure 15.3 shows the cavity wall with these details (discussed in the two previous chapters) to identify the dead and imposed loads together with the wind loads. Inspection shows that the internal skin takes the dead and imposed loads. No strength requirements for these loads must be considered for the external skin, so the facing brick should be sufficient. Durable bricks will have reasonable strength, so the external skin should be able to support its self-weight and deal with any local crushing from the lintels over openings. If large openings are included in the design, the strength of the brick may become an issue.

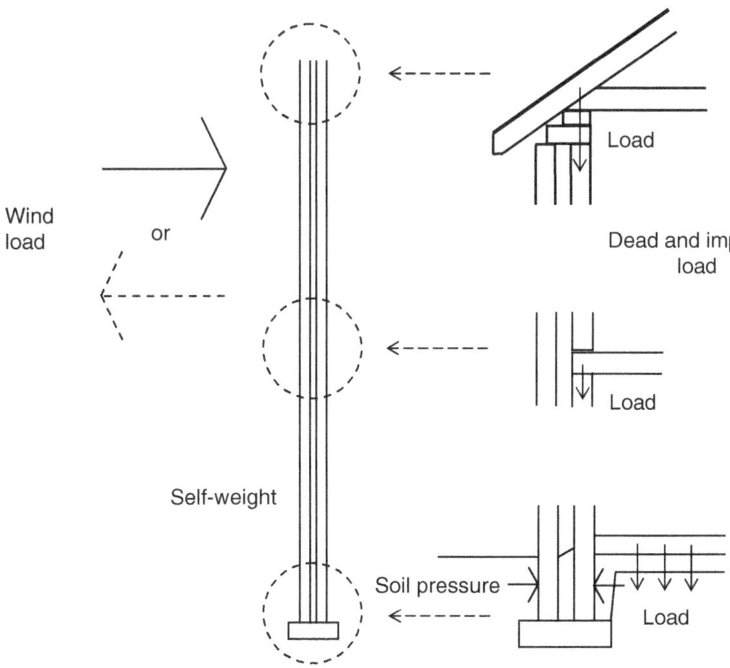

Figure 15.3 Loads on the external cavity wall.

One of the key aspects of the specification of the block for the internal skin will have to be strength. The most highly stressed areas of the internal skin will be between the ground-floor windowsills. Given a reasonable distribution of windows in both size and spacing apart and taking the skin to be 100 mm thick, it should be possible to use aerated concrete insulating blocks at 2.8 N/mm^2 (these may have a minimum thickness of 115 mm) for dwellings up to two stories high. These blocks are normally more expensive than blocks made with lightweight aggregates that are denser and therefore offer less insulation value but more thermal capacity. If high thermal capacity is required, dense concrete blocks may be a better choice as they are less expensive, and certainly strong enough, but will have poor thermal insulation properties. It is therefore best to leave the final specification of the block until the thermal behaviour of the wall is considered.

The structure – overall stability

Returning to Figure 15.3, another observation would be that the cavity wall is slender, making it fundamentally unstable. Given that there are adequate foundations, two other instabilities can be identified. The first is overturning by wind forces and the second is buckling of the internal skin under compressive forces from the dead and imposed loads.

The mortar jointing and the pattern of bonding in both the bricks and the blocks ensure that the wall acts as a panel. However, as presently discussed, the external and internal skins are independent. Wind loads on the external skin are not transferred to the internal skin and the external skin does not help the buckling of the internal skin. Mutual structural support is required. This is normally provided by wall ties (although under most conditions they will probably act as struts). The form of these ties and the spacing required are shown in Figure 15.4, ranging from the simple butterfly type through to the more modern vertical twist type, which can also include a disk to restrain the thermal insulation (not shown in the figure). The width of the cavity will determine the type of tie that can be used, with the vertical twist being the only form used in wider cavities over 100 mm. On all types of wall tie there is a drip, a device to stop water travelling horizontally, bridging the cavity across the wall tie and causing damp in the inner skin. Ties would now normally be specified in austenitic stainless steel, although galvanised steel and plastic forms have been used. With

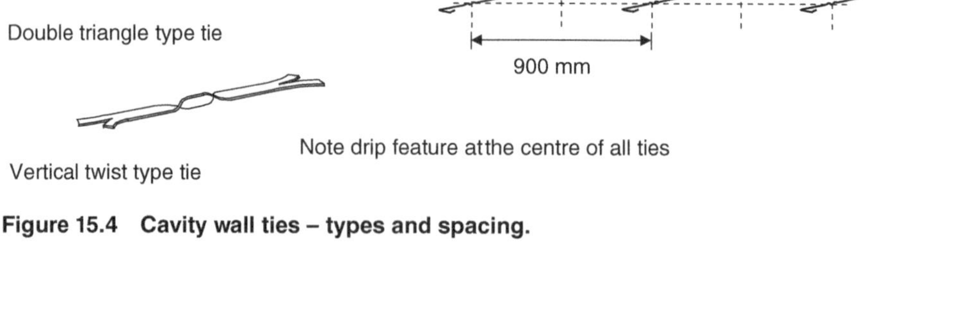

Figure 15.4 Cavity wall ties – types and spacing.

the introduction of wall ties the wall now acts as one panel for both overturning and buckling but is still unstable.

Any return wall will act as buttresses to resist overturning. If the walls of the house form a cellular pattern on plan, they will receive mutual support. If the walls are predominantly in one direction as in a form known as cross-wall construction, piers will have to be formed, normally at both ends of the cross wall. This leaves one potential area of wall vulnerable to overturning. The wall forming any gable ends will not benefit from the return walls to give stability against the wind forces. A chimney may form a pier, but in the absence of any masonry support the wall should be strapped to the roof to gain additional stability. Straps should be provided at the top of the gable wall and then at not more than 2000 mm centres down the slope of the roof. It is also necessary to use the support of the upper floors to reduce the effective dimensions of the wall against buckling. This is achieved by providing joist hangers with restraint or strapping. Straps should be placed at not more than 2000 mm centres for walls over 3000 mm long.

Rising damp

The cavity provides resistance to penetrating damp from rain falling directly on the surface of the wall. The materials chosen are permeable, so moisture can enter the building by rising from the ground where capillary action within the materials will draw water into the bricks and blocks, creating a path for damp. The solution is to provide a damp-proof course (DPC) across each skin. In a cavity wall these DPCs are independent and can therefore be at different levels. For the external skin, the DPC must be at least 150 mm above ground level. This protects the wall from rain falling on the ground, which will splash back against the wall. For the internal skin, the level depends on the internal floor level and the type of floor construction. For a suspended ground floor the DPC must be lower than any part of the floor construction that would be vulnerable to damp, while for a ground-supported slab the DPC must be continuous with the damp-proof membrane (DPM). This is shown in Figure 15.5.

A variety of materials can be chosen for DPCs. In the past, materials such as slate and even bricks with very low permeability known as engineering bricks have been incorporated in the wall, but thin, flexible sheet materials that can be laid in the bedding joint are now used. Metals such as lead and copper can be used but are expensive. The less expensive bitumen DPC, and the polymeric (bitumen polymer and pitch polymer) sheets, are the most widely used. The polymeric sheets are also used for other DPC applications such as

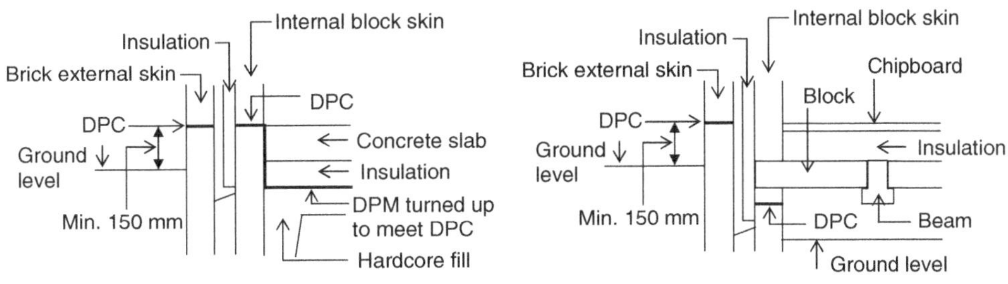

Figure 15.5 Masonry cavity wall – DPC arrangements.

cavity trays, which will be discussed later in this chapter. The polymeric materials provide a tough material that will not extrude under load, give good adherence to the mortar and have good durability even when bent to form gutters. Although naturally black, it is possible to obtain the polymeric material in mortar-matched colours to reduce the visual impact of the black line of the DPC in the mortar joint.

Thermal performance

With concerns for energy saving regarding winter heating in houses, the initial focus for thermal performance has been on insulation properties to resist the passage of heat. The focus is on the elemental U value, the thermal transmittance of the wall, as an individual element. Still air provides good insulation, so the unventilated cavity not only gives the weatherproofing that is its main role; but also contributes to the overall thermal resistance of the wall if it is at least 25 mm wide. This can be enhanced by providing a low-emissive surface facing the cavity to limit radiation transfer. It is also possible to increase the resistance of the wall by careful specification of the block.

Concrete blocks vary in their density. It has already been established that higher densities give greater strength and thermal capacity but lower insulation values. Selecting a low-density block can significantly affect the insulation value of the wall but still provide sufficient strength for a two-storey house. However, even with the cavity and the low-density blocks, the wall will not have a thermal transmittance or U value sufficient to meet either the current regulations or the expectations of wider environmental concerns.

It is possible to increase the thickness of the most effective material (in this case the block, as increasing the width of the cavity beyond 25 mm will have little effect on thermal performance) or introduce a layer of material specifically for its thermal insulating properties. The most common solution is to include an insulating material in the cavity. Either partial fill (still leaving a cavity) or total fill solutions can be achieved with bats or boards, and total fill solutions can be achieved with loose materials or *in situ* processes installed after the walls are completed. Many combinations of block and insulation within the cavity are available and which meet elemental U values required to satisfy the requirements of the building regulations. These can be found in current construction technology books, see for example *Barry's Introduction to Construction of Buildings*. Approaches that go beyond the current thermal requirements of the Building Regulations can be found in guidance on passive house construction.

The external wall is only one part of the construction through which heat is lost. Door, window, roof and floor elements all have different U values, and it is possible to trade off the high insulation value in one element for another if the overall heat loss from the house will not be affected. This leads to the specification of target U values that can modify the required performance of the wall.

Although the performance for the wall was initially set in terms of resistance to the passage of heat with its focus on limiting fabric losses, the wider purpose of insulation is the conservation of energy. Walls and their insulation are only one aspect of energy saving for the building. Elemental U values are still important but are increasingly only part of the scheme to limit the energy use in a house. Aspects of ventilation rates, water heating, internal gains and solar gains are considered along with fabric losses in energy-rating schemes that look at the energy consumption of the building. The external wall may now have performance requirements beyond thermal insulation. The openings will influence ventilation rates, as will the detailing of joints and junctions where airtightness is now a consideration. The thermal capacity of the fabric can be used to harness the heat gains to reduce the overall heating requirements for the house. As thermal capacity becomes significant so the position of the insulation in the wall must be considered.

To understand why this becomes significant it is necessary to consider the wall not in steady state but to see the temperatures on either side of the wall changing. The effect of heat gains raising the air temperature will be limited by a wall with a high thermal capacity, because the wall will take a longer time to heat up towards steady state. Further, when air temperature starts to fall, the heat stored in the wall will be given back to the air, delaying the need for additional heat inputs. A similar cycle can be used for summer cooling in larger commercial buildings.

The position of the insulation will also affect the condensation risk. The thermal behaviour of the wall for condensation risk, response times and thermal capacity issues are explained in Chapter 8. The other major aspect of the wall resisting the passage of heat is the concern for thermal bridging. With masonry construction, this occurs at openings and junctions with the floor and roof. Full details of this can be found in *Barry's Introduction to Construction of* Buildings.

Size of the cavity

It is now possible to return to confirm the size of the cavity that was initially set at 50 mm wide. It was suggested that to act as a barrier to penetrating damp the cavity must be wide enough to be kept clean, and this will be influenced by both the inclusion of insulation and the production process. The cavity should be at least 25 mm if it is to achieve the thermal insulation potential of the still air providing good surface resistance on both sides of the cavity, so the 50 mm is satisfactory for the thermal function.

The influence of the production process can now also be identified. The process will involve bringing up the inner skin while protecting the cavity from mortar dropping as the work proceeds. If insulating boards or bats are to be installed in the cavity, these are then placed and restrained against the inner skin (quilts will not be stiff enough to give a stable face inside the cavity). It is then possible to lay the external bricks. Two aspects about laying the external brickwork now need to be questioned. How easy is it to clean the mortar that squeezes out from between the bricks as they are tapped to level and how accurately can the width of the cavity be maintained? While it was easy to clean the mortar from the blocks forming the inner skin, the bricks are forming the cavity, and the bricklayer must clean the mortar with the trowel from inside the cavity. If the cavity is too small, there is the danger that the mortar protrudes into the cavity and cannot be cleaned or falls into the cavity where it will build up at the bottom of the wall, at window heads and even catch on wall ties, providing a damp path from external to internal skin. This will be aggravated by any narrowing of the cavity to keep the external wall in line, given the accuracy that was achievable in laying the internal block skin first. Experience has shown that it is only fair to the bricklayer to have a cavity around 50 mm to keep the cavity clean to achieve the insulation value and eliminate the risk of penetrating damp.

Many options for insulating the wall involve the cavity in different ways, some of which have been introduced in the section on thermal performance in this chapter. As an example, a wall could be specified with a cavity with partial fill insulation of 50 mm and a clear cavity of 50 mm, giving an overall distance between the inner and outer leaf of 100 mm. Given that the wall tie is often in compression acting as a strut, some thought must be given to buckling as the distance between the skins increases. Wall ties were originally designed for clear 50 mm cavities with no insulation. Ties must be embedded in the masonry joint around 60–65 mm, and they must be stiff enough to resist buckling across the free length in the cavity. Ties for 100 mm wide cavities, or wider, have therefore to be longer and thicker. This requires the correct type of wall tie to be specified to suit the overall cavity width. Guidance on the correct size of wall tie for the specific cavity width can be found in manufacturers' information.

The wall so far

The variety of options for thermal insulation, its position and type of material, with its thickness in conjunction with the specification for the block and its influence on the cavity, have already been identified. Appearance and exposure conditions lead to a variety of performance needs that influence the choice of the external skin. Making a single choice for this case study is therefore not possible, but Figure 15.6 shows one option adopting the partial fill cavity insulation given as an example in the paragraph above. This wall has been chosen to achieve weathering, structural and thermal performance. As specified, it will provide a good level of sound insulation as it has both mass and a discontinuity with the cavity, and the material chosen will give good fire resistance. The basic structural and environmental functions will be fulfilled by this construction.

Internal and external finishes

The external appearance has been set by the choice of facing brick, although detailing and the use of more than one brick and the colour of the mortar or the use of renders can give design options. The shape of the front of the mortar joint is also sometimes used as a design feature, but it also has a significant effect on the weathering and durability of the wall. These

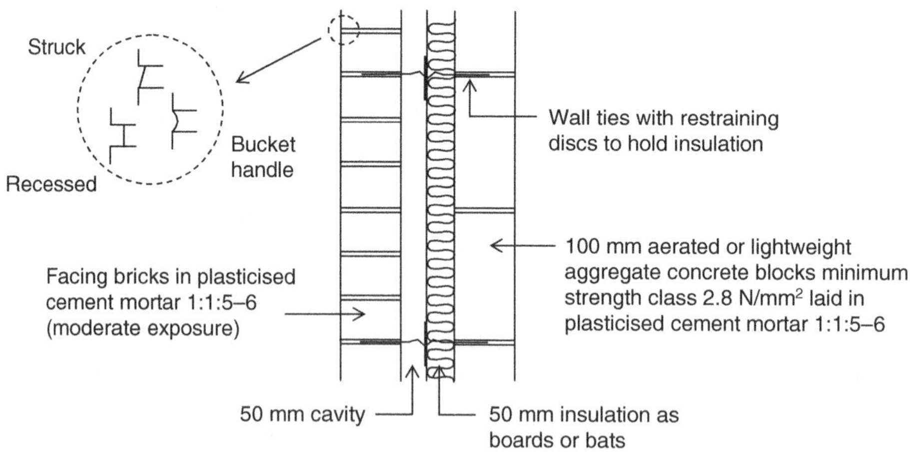

Figure 15.6 Masonry cavity wall – basic specification.

alternative joint finishes are shown in Figure 15.6. The struck or weathered joint is a tooled joint (it has the surface smoothed after the mortar has started to cure), which promotes the run-off of water and protects the brick, while the recessed joint has the opposite effect, leaving the soaked edge of the brick vulnerable to frost attack. The struck or weathered joint takes time and skill, and therefore the more straightforward, tooled bucket handle joint is often specified. A fourth joint finish, the flush joint produced as the work proceeds without tooling, does not have good resistance to rain penetration and may not produce a good edge on uneven or open-textured bricks.

Internally, concrete blocks can be specified for a fair-face for direct painting, but this is unlikely to be satisfactory for a house. The options will usually be between a wet process, probably lightweight gypsum plaster, and a dry plasterboard system. The blockwork provides the background or substrate for the finish. It is necessary for the background for the finishes to provide:

- ❏ Sufficient stable support
- ❏ Suitable fixing opportunities
- ❏ Compatible flatness tolerances
- ❏ Matching for future movement
- ❏ Avoiding deterioration mechanisms

If plaster is used, the background characteristics of the blockwork will be satisfactory for the applications of two-coat work. The wall is clearly stable, as it is part of the structure. The density of the blocks is significant in that it determines the suction during plastering that makes the bond that provides the fixing between the block and the plaster, and this will influence the specification of the plaster. The tolerances on the wall should be well within the capability of the first coat of plaster, normally about 12 mm thick to bring the surface to a flat plane. The final skim coat of plaster will then provide a tough, smooth surface for decoration. Plaster shrinkage should not be sufficient to break the bond with the wall, although shrinkage cracks may appear at the top and bottom of the wall. If blocks are inert, there should be no interaction causing deterioration of either block or plaster. However, if the blocks are laid and finished before they have completely cured and continue to shrink, cracks may appear in the plaster.

The plaster surface, while tough enough for most domestic situations, takes only limited-impact loads and is not suitable for external exposure. If either of these conditions is experienced, renders should be used. Based on sand and cement with a plasticiser, traditionally lime, these finishes can be used externally even in severe exposure conditions with the appropriate specification. If renders are used externally then, for economy, a concrete block could replace the brick external skin. As with plasters, the density of the block chosen would determine the strength of the background. The first coat of render would have to be specified to match the strength of the block, with the final coat(s) determined by the exposure conditions. Finishes on external renders can vary from smooth trowelled finishes that can be painted to thrown aggregate finishes using decorative aggregates to provide a weathering surface.

Dry finishing systems for internal walls will most likely be based on the use of plasterboard. As a background the blocks are not sufficiently flat to apply the boards directly to the wall. The wall is, however, stable and ready to accept a range of fixings. Two options are possible: to provide timber battens levelled to receive the board to be fixed with nails or screws, or to use dabs of a plaster-like material to both fix and level the board to the wall. With both these systems the boards still must be jointed. While square-edged plasterboard (fixed showing the grey face) with the joints reinforced with scrim can then be skimmed with plaster, the use of tapered edge boards allows the joint (and screw holes) to be finished without having to plaster the whole wall.

Interaction with other elements of construction

The interactions with the floor and the roof in both load transfer (roof/floor to wall) and the provision of stability (wall by roof/floor) have been identified. Services also interact with the wall and therefore may influence the specification and detailing. There may have to be some hiding of wires below the plaster in chases in the internal skin, but most pipes will be surface mounted and then boxed in if there is a wish to hide them. Radiators and other components will be fixed to the wall, demanding some consideration of the strength of the wall and the ability to identify suitable fixing devices to take the loads involved. This will also be true of the need of most occupiers to have shelves and cupboards fixed to the wall, particularly in kitchens. A range of fixings is available for a variety of block walls and even the low-density blocks have the capacity, with the correct fixings, to take these loads, so the specification so far suggested will meet these performance requirements.

Life cycle considerations

Choice of the external material with concern for the exposure conditions to ensure durability has been discussed, as has the choice of the internal finish. The insulation and the internal skin should remain dry, so, in the absence of infestations in the cavity damaging the insulation, there should be little deterioration of these materials, although full cavity loose fills may compact with time. Maintenance requirements should be minimal if the external bricks are laid in cement mortar. Modification and adaptation will have to be undertaken with care, as the wall is a loadbearing structure. Any changes to the external wall may also have difficulty in matching bricks at some time in the future, but internal finishes are easy to reinstate.

Consideration of the recycling and or reuse of the masonry components at the end of their life indicates opportunities for recycling as fill and reuse in new projects if the bricks and blocks are still in good condition. Most facing bricks will have a long service life, and although this varies depending on the qualities of the bricks, the detailing of the building, and the quality of the jointing, it is not unreasonable to have a service life of over 100 years. Many buildings may be altered and disassembled long before the external brickwork has deteriorated to such an extent that it cannot be recovered and reused. The ease of recovery of individual bricks will depend on the type of mortar used and the strength of the bond. It will be much easier to dismantle a brick wall that is constructed using lime mortar compared to one that has used cement mortar because the lime is comparatively soft and can be removed without damaging the bricks. This is more challenging when concrete mortar is used, which may be harder than the brick to which it is attached. The thermal insulation should be easy to separate from the wall assembly as the wall is dismantled. The insulation is only restrained in position, although it may have mortar from the blockwork stuck to the back that may be difficult to remove. The simple inclusion of a paper layer between the insulation and the block so that the insulation does not bond to mortar may be an inexpensive way to ensure the insulation could be reused. Metal wall ties could also be recycled and may be in sufficiently good condition to be reused.

15.5 Paradox

As the masonry cavity wall has evolved to satisfy ever stringent thermal performance requirements it has changed from a simple construction to a complex assemblage of materials. In most cases the cavity is now filled with insulation to create a sandwich construction. We continue to use the term cavity wall even when there is no cavity. This raises questions about whether a 'cavity' wall is suitable for our case study house. It could be argued that we should stop using the term cavity wall if there is no airgap, and instead use the term sandwich wall or composite wall. Regardless of what terminology we choose to use, it appears that the masonry cavity wall needs a rethink. It has, one could argue, become too complex in response to ever more demanding thermal performance requirements. This makes the construction of the cavity wall on the building site challenging to achieve the required performance. It may also pose challenges for future disassembly and reuse of materials at the end of the building's service life.

Do we need to use cavity wall construction?

The fundamental question to ask is do we need to use a masonry cavity wall? Although this form of construction is very familiar, it does not mean that we should keep using it if it has outgrown its original function. The original function was, as discussed earlier, to keep moisture from penetrating the inside of the building. That function remains; however, the requirement for thermal insulation and airtightness provides the stimulus for different solutions. Some of these alternatives may be simpler to construct, for example using offsite production methods, and some may have economic benefits and/or better environmental credentials, such as Passivhaus construction principles. For the purposes of analysis and choice we need to start by asking what we need the wall to do and why.

Analysis

Analysing a typical masonry cavity wall there are a few dominant features (see Figure 15.6 and *Barry's Introduction to Construction of Buildings*). The external leaf is usually facing brickwork or stone, or blockwork protected by render. The cavity contains the thermal insulation (full fill), and in some cases an air gap (partial fill). The inner leaf comprises blockwork, which is usually finished with plaster.

The external leaf has not changed much over the years. Bricks remain similar in appearance and performance even though the production methods have become more sophisticated and less energy hungry. Specifiers may choose to use locally produced bricks to mitigate the carbon produced in transporting the bricks. The mortar tends to be cement mortar, not lime mortar. Much of the analysis and choice is concerned with the properties of the brick and its appearance, namely its colour and texture. Stone tends to be sourced locally, although it has become common to use artificial stone. This is a concrete based product that is mixed with stone to produce a product that resembles natural stone. In some cases, it may be possible or desirable to use reclaimed brick or stone.

The internal leaf will be constructed using lightweight concrete blocks. The blocks have been manufactured to increase their thermal resistance while retaining their loadbearing properties. Choice will be made according to the required loadbearing capacity and the required thermal resistance of the block. Choice may also be influenced by the carbon footprint of the block, with manufacturers marketing some of their products as 'green' blocks to reflect improvements in the environmental performance of the block.

The thermal insulation is usually chosen based on its thermal resistance and its thickness. The thicker products will take up more space in the cavity compared to the thinner products. The thicker wall section will also require a wider foundation and longer and stronger wall ties. While this may not be significant concern for individual houses, the large housebuilding companies will be keen to maximise density of housing on a site, and wall thickness is a consideration. Take for example the case of a large development. Product A and Product B are thermal insulation products that provide the same U value for the cavity wall. Product A results in a wall that is 300 mm think. Product B results in a wall that is 350 mm thick. The additional 50 mm thickness per wall will result in a house that is 100 mm wider (two external walls), but with the same internal dimensions. The larger footprint of the house, when multiplied ten times, results in one metre extra space compared to the thinner product. This may not sound significant but multiplied many times over a large development site and there will be fewer houses built compared to the option with the thinner insultation material. This impacts profit margins, and hence thinner thermal insulating products tend to be favoured over the thicker ones.

Thickness and thermal performance should not be the only considerations. Thermal insulation also needs to be analysed for its environmental impact. The choice of products is extensive; however, they can be classified into four main groups:

- Inorganic materials derived from minerals, typically stone wool (rock mineral wool) and glass mineral wool products. Also included are vermiculite and aerated or foamed glass and aerogels.
- Organic fossil fuel derived products. These are derived from oil and include expanded polystyrene (EPS), extruded polystyrene (XPS), polyurethane (PU), polyisocyanurate and also phenolic foam products.
- Organic plant and animal derived products. These include sheep's wool, hemp, hempcrete, cellulose, flax, cotton, cork, woodfibre, woodwool and compressed straw.
- Newer products such as biopolymers, nanocoatings, phase change materials and advanced foams, that aim to offer space saving qualities compared to some of the other options.

Thermal insulation products, in combination with the overall wall construction also contribute to the wall's hygrothermal performance, its performance in the event of a fire, and other factors such as acoustic performance and indoor air quality. This applies to the masonry cavity wall example and the alternative construction solutions noted below. It also applies to the thermal upgrading of existing walls. Analysis and choice are, therefore, not as simple or as straightforward as one may initially think by looking at the cross section of a wall in construction technology books.

Alternatives

When considering alternative solutions to the masonry cavity wall it is essential to consider the function of the wall in relation to net zero and wider environmental impact. There are other forms of construction that can be used to fulfil the functions of an external wall. These include solid masonry walls and framed construction using timber or lightweight steel. Solid loadbearing walls will be constructed on the site with the thermal insulation fixed to the inside or outside of the wall and then protected with a cement or lime-based render. Some solutions involve the use of organic bio-based materials, for example, straw bale walls with hemp-lime render finishes. Framed systems combine varying degrees of prefabricated elements using timber, steel, or concrete. This includes approaches such as structurally insulated panels (SIPs) and insulated concrete framework (ICF). These are described in *Barry's Advanced Construction of Building.*

Reflective exercises

A conscious decision has been taken to not include any drawings, sections and details, to allow readers to develop their own conceptual details in relation to the design of the external wall. Continuing with our case study building, how would the details look if we embraced some of the thinking outlined in Section 15.5.

Exercise 1. Try drawing a section of an external wall from first principles without relying on existing details to be found in the construction technology books. Think creatively, do not be bounded by convention.

Exercise 2. If Exercise 1 proves to be too challenging, try taking an existing detail, for example one of those provided in this chapter or in *Barry's Introduction to Construction of Buildings* as your starting point. Then analyse the chosen detail and explore what could be changed to create an external wall design with better environmental credentials. How would you justify your decisions?

Openings – The Door and Window Performance Dilemma: Chapter 16 AT A GLANCE

What? To provide all the functions of an external envelope it is necessary to install windows and doors. To accommodate these components openings are formed in the wall, which must maintain the performance of the wall once occupied by the door or window. The windows and doors are usually designed and manufactured by specialists and require only detailing to fix and weather the units into the openings. However, the size of the opening, the performance and the operation of the components and the materials from which they are made must be chosen as part of the overall environmental design for the house. Given that it is very challenging to match the thermal performance of the doors and windows with that of the wall, the question arises as to whether there is a different choice.

Why? Windows and doors offer a number of performance criteria that cannot be achieved with a wall. Doors provide the means of entry and exit from the property, in addition to privacy and security. Windows provide a means of allowing natural light into the building interior, in addition to the provision of ventilation when open. Both windows and doors assemblies need to satisfy the performance criteria for the wall as an entity, for example in relation to fire resistance, the passage of heat and the transmission of sound. They also add to the appearance of the building and therefore are important design considerations. If we were starting from first principles and wanted to design openings in walls to achieve very similar thermal performance to the wall, how would we go about it? Could we use layers instead of sophisticated technologies, for example, lobbies (with an external and an internal door) and double and triple window systems instead of highly insulated doors and windows with sealed triple glazing units? These solutions may take up more space compared to the more common solutions, however, with careful design these low-tech solutions may present more sustainable solutions compared to the high-tech solutions.

When? Windows and doors are usually installed into the openings once the main structure is complete. This helps to reduce the likelihood of accidental damage to the components during construction. Alternatively, they can be built into the walls as the house is constructed. In most cases this usually involves the fixing of the door or window frames, with the doors and glazing components being fixed later in the process. If we are taking a different route to the overall design of windows and doors, then this needs to be addressed at the conceptual design stage and through into the detailed design.

How? The majority of doors and window assemblies are manufactured in factories and are chosen from manufacturers' catalogues. In some cases, it may be preferable or necessary to design and fabricate bespoke doors and windows, for example in refurbishment projects to suit existing opening sizes. The door and window frames will be fixed to the wall, usually by screwing the frame to the wall, with the door and windows than hung in the frame.

16 Openings: The Door and Window Performance Dilemma

The wall considered so far does not provide access, light or ventilation, which will be required, and which will vary depending on the design of the house. When we form a hole in an external wall for a window or door, we compromise the integrity of the wall assembly and its overall performance. The wall has a similar, yet different, function to the window or door. How then do we get the performance of windows and walls to the same level as the main building envelope? This has challenged designers since the early days of performance criteria. In this chapter we challenge the reader to re-think the performance of doors and windows for a small domestic dwelling. Do we have to accept the normal way of doing things? Are there alternatives?

16.1 The hole in the external wall

Access, light and views and ventilation, will be required in an external wall. All buildings need some form of access, and with a few exceptions all buildings require windows and/or vents. These components cannot take loads and will have less thermal, sound and fire resistance compared to the wall into which they are being placed. They also create a safety and security risk and may compromise privacy. This is normally resolved by creating an opening to maintain the integrity and performance of the wall while designing independent components for the doors and windows to complement the wall and ensure the function of the wall as a single entity. To this must be added the need to provide fixing and jointing of the components into the opening. While the size of the opening will be dictated by the design requirements of the door and/or window, the basic detailing of the opening for the wall to the house is unlikely to vary.

The terminology used in openings is shown in Figure 16.1. The detailing of the head, jamb and sill (threshold) will each need to be considered separately but are linked in the need to keep the depth of the reveal and soffit constant around the opening. The head will need to ensure that the loads are transferred to the jambs. The jambs will then need to be stabilised as they take the additional compressive forces. The sill has no structural requirements as it can be formed on the wall below, but the weathering requirements and the exposure conditions are more acute at the sill. This must be achieved with an accuracy to ensure the doors and windows, now almost certainly components made offsite, can be fitted and weathered into the opening. While some packing may be tolerated below the sill, the fit between the jambs must be accommodated in the fixing and weathering details of the component to the reveal.

Barry's Construction Technology: Analysis and Choice, Third Edition. Stephen Emmitt and Tony Bryan.

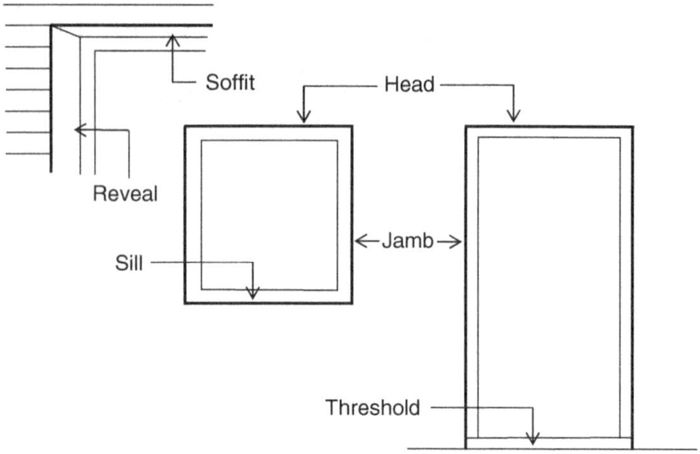

Figure 16.1 Openings in walls – terminology.

Detailing the head

Carrying the load across the opening is fundamental to the detailing of the head. In the past arches have been used because the basic material of the wall is weak in tension, so, with the basic structural action of the arch being in compression, the arch allowed the same wall material to be used to span the opening. With the greater range of materials now available the lintel (a small beam with a bending action) will almost certainly be used.

If the lintel is based on a rectangular cross section carrying both skins as a simple beam, the top of the lintel will collect water that can run back towards the inside of the wall, and the front of the lintel may need a finish to improve the external appearance and/or durability. The rectangular section has been used in the past for stone, timber and reinforced concrete lintels. All these materials will create a thermal bridge if the insulation is provided in the cavity, although this would not be the case if internal or external insulation were used.

While all these difficulties can be overcome, developments in pressed steel and lightweight concrete provide economic solutions that take all these concerns into account. Two alternative details of the head of an opening showing a steel lintel and a lightweight concrete lintel are given in Figure 16.2.

These solutions recognise that for a two-storey wall the loads are concentrated on the internal skin and the need for strength to support the outer skin is limited. In the lightweight reinforced concrete solution the basic rectangular section remains to support the loads from the internal skin, although it is now shaped to form a cavity tray to direct water away from the inner skin. The gutter emerges at the steel shelf angle that supports the external skin, and weep holes formed in some of the perpendicular joints between the bricks help remove the water more efficiently. It is now possible to bring the brickwork across the opening to visually give continuity to the external finish and protect the lintel.

The pressed-steel solution gains its strength not from a solid material but from shaping sheet steel. The weathering details and appearance remain the same as the composite concrete and steel shelf angle solution. The major advantage is the low weight of the lintel, an advantage in the production process. The hollow section can be filled with insulation

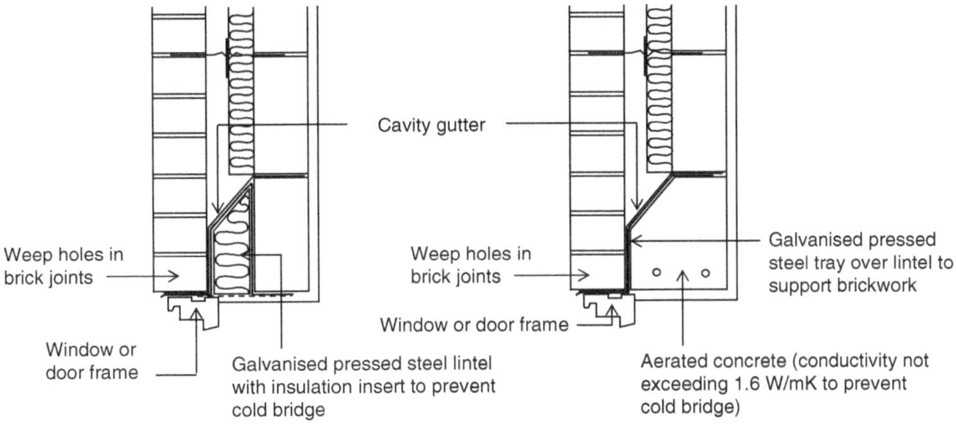

Figure 16.2 Openings in walls – alternative head details.

to reduce cold bridging and some means of providing a background for the internal finish must be incorporated under the lintel. The steel must be protected from corrosion with a galvanising layer, but even this may be suspect over time in exposed conditions, particularly with the risk of damage during the production process. It is possible to specify polyester-coated lintels to provide extra protection.

One detail that cannot be seen from the cross sections is that the cavity tray, forming the gutter from sheet DPC (damp proof course), should either extend beyond the lintel to minimise the risk of moisture running back off the end of the gutter and across the underside of the lintel or use stop ends to the gutter to prevent the water running off the ends and wetting the cavity insulation.

Detailing the jamb and the sill

The detailing at the jamb is relatively simple, and two options are shown in Figure 16.3. The first shows the return of one skin to close the cavity and then providing a vertical DPC to eliminate damp penetration across the jamb. Which skin is returned will determine how deep the reveal is, as the frame of the window or the door will be located to cover the DPC, ensuring no moisture reaches the internal finishes or the back of the frame if this is against the external skin. Returning the block (detail shown) will reduce the thermal bridging effect but will leave the frame close to the external face. The reveal will be small. This means that the sill on the frame can be detailed to hang over the wall without the need for a sub-sill but brings the joint between the frame and the opening into a more exposed position. To give the joint better protection, the outer skin can be returned to gain a deeper reveal. A sub-sill will now be required, and the brick return creates more of a cold bridge, although insulated DPC can help reduce the effect. The second option is to use a cavity closure. This hollow plastic box section can be filled with insulation to reduce the cold bridge, act as a DPC and in addition can provide a fixing for the frame.

Structurally the edge of the wall must take the load from the lintel but has lost the stabilising effect of the bonding, and this makes the jamb subject to buckling. The return skin or the cavity closure provides support if the skins buckle towards each other, but it provides

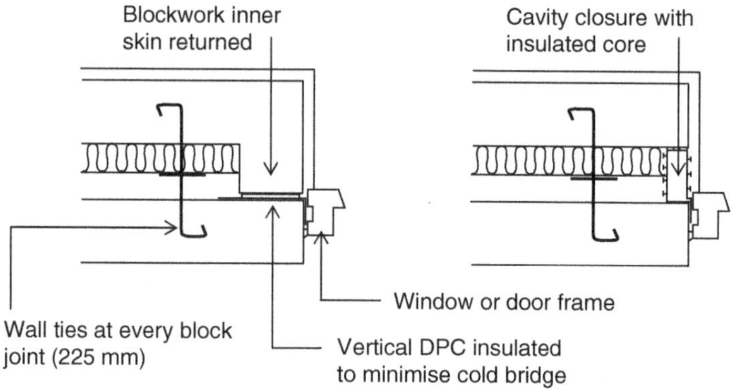

Figure 16.3 Openings in walls – alternative jamb details.

no resistance to them moving apart. To provide this resistance, wall ties are placed in every block joint adjacent to the returned skin. The cavity closure can incorporate ties that attach to the box section.

The detailing of the sill, with and without a sub-sill, is shown in Figure 16.4. The major concern is to shed the water that runs off the impermeable window or door surface away from the wall as effectively as possible. The top surface of the sill should be sloping, the lip should overhang the wall and the underside of the lip must be provided with a drip. This amount of water around the sill raises the concerns of durability so the material of the sill needs to be chosen with care. Moisture penetration is an additional concern, particularly at the ends of the sill at the reveal and under the frame, where wind-blown rain will be held and may even be under pressure rather than running off down the slope provided on the sill. If the cavity is closed in the sill detail, a DPC will be required to protect the internal finishes, although bedding the sill on a DPC is common practice to protect a timber frame sitting on the external skin. Care must be taken to ensure that the edge of the DPC is not laid forward of the line of the front of the wall, or the drip may be breached and water drawn back under the sill. Cavity closure sections can also be used in the sill detail.

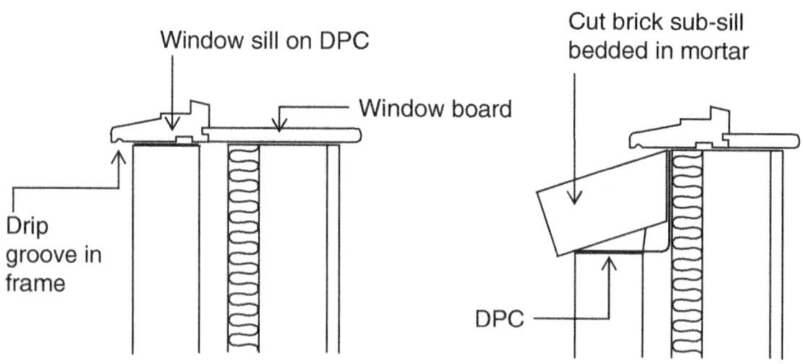

Figure 16.4 Openings in walls – alternative sill details.

16.2 Doors and windows – the components in the openings

It is not the intention to consider the details of the components themselves. This has become a specialist area of expertise, where the manufacturing of sections and the factory assembly of the components must be understood as well as the component design aspects such as strength, weathering, safety and security. All these are beyond the scope of this text. However, the design of doors and particularly windows with the joint and fixings into the openings needs to be thought through, particularly as concerns for sustainable development are leading to a re-evaluation of energy saving by considering the behaviour of the house as one entity. This includes a consideration of the orientation of the façade, the area or percentage of glazing, the airtightness of the joints and the shape of the windows, as well as the specification of the materials involved. While this has a significant impact on the look of the house and the views experienced by the occupants, it also has an impact on the energy efficiency of the house.

Given the predominant energy used in houses during winter heating conditions, any contribution that can be made to reducing heat loss or increasing gains needs to be considered. Windows have the potential to do both. Solar gain from south-facing façades can be usefully used for many days in winter, where direct radiation from the sun will pass through the glass and heat objects in the room, adding to the heat gains. However, the glass is poor at retaining the heat from the air in the room, increasing heat loss. Heat loss can be considerably reduced not only with double (or triple) glazing but also by the choice of glass with coatings or gas-filled cavities that can reduce U values of double glazing almost to that of triple without the expense of the more complex frames and opening casements.

Using glazing for heat gains is only effective for façades facing in a southerly direction. In these façades larger areas of glazing can provide a net gain, but this is not true for other façades. It is better to limit the area of glazing as heat gains are not possible and the wall construction can provide much higher U values to reduce heat loss compared to the windows. Double glazing with coated glass can still make a significant contribution to reducing heat loss.

These choices to reduce winter energy use have consequences for both lighting in the rooms and heat gains during the summer. Light from the south can be very bright and cause glare, so larger windows may create discomfort or even disability during the summer, requiring the use of blinds or curtains. Similarly heat gains will be greater, possibly requiring shading. Conversely, while north light is good quality and more even, the smaller windows may make rooms dim during the winter, requiring artificial lighting and adding to energy use.

Currently the number of days when summer heat gain or poor lighting are likely to be of concern in the UK is limited as for many households daytime occupation is limited (unlike commercial property where this is a major problem, particularly with heat gains from equipment and occupants). For the days when conditions are uncomfortable, ventilation by opening windows is the main mechanism offered to householders to control the internal temperature. If occupation patterns change or vary considerably, this may have to be reconsidered. This also applies to changes in our climate, and designs including light-coloured walls, screening and verandas that are used for houses in hot climates may be considered for some designs in the UK. Perhaps high thermal capacity construction with night-time purging being used in commercial property in temperate climates will be adopted for house construction to help temper the internal environment.

Having considered shape and size along with any additional components such as blinds or shading, the choice of materials must be considered. Glass has been discussed but not the choice of material for the frame. For domestic buildings, timber (soft and hardwood) or plastic (unplasticized polyvinyl chloride, uPVC) are the main choices. Aluminium is also specified, but it has a higher capital cost than timber or plastic. Steel has proved to have a high maintenance cost to control corrosion. However, galvanised, polyester coatings and stainless steel overcome these problems, but are not often considered for housing given the viable options of timber and plastic. Aluminium can give slender frame sections and be produced with coloured anodising if this is required for appearance. The choice between these frame materials will include an analysis of the need for maintenance, potential service life, ease of replacement and potential recycling and reuse options. Timber (particularly hardwoods) can have an impact on habitat and global climate, while plastics have pollution, energy and environmental considerations associated with their manufacture and subsequent disposal.

Having chosen the size, shape and material for the windows and doors and had them manufactured to give strength, be weather-tight and secure, it is necessary to consider how they are to be fixed and weathered in the opening.

Windows and doors can be built in as the work proceeds, being fixed back to the brickwork with frame cramps or secured to the cavity closure with screws. They can also be installed after the brickwork is complete by screwing through the jambs. If they are fixed after the brickwork is complete, the bricklayer has lost any direct control on tolerances to maintain the gap between frame and brickwork where the appearance and weathering functions must be achieved. This will have to be replaced either by careful measurement of opening size, square and plumb or with the use of temporary frame forms. Packing will then have to be used between frame and brickwork to ensure the jambs of the frame are not distorted as the window is secured. However the frame is fixed, weatherproofing is normally achieved with gunned mastic, although consideration should be given to placing the window towards the back of the reveal to give greater protection, particularly in exposed conditions. Some of the fixing methods make it easier to replace the window than others, which may be a consideration when thinking about the life cycle of the building and the need to replace windows after a given timescale.

16.3 How could we do things differently?

So far in this chapter we have outlined the basic construction that relates to the hole in the wall. These are filled with doors and windows that are usually selected from manufacturers' catalogues of standard products. Very few windows or doors are bespoke designs in domestic new buildings. The most variation is reserved for retrofitting and upgrading of existing buildings where the opening sizes may require bespoke sizing of window and door frames. Given the pressing need to address the climate emergency, how could we do things differently for new build properties? If we were starting from first principles and wanted to design openings in walls to achieve very similar thermal performance to the wall, how would we go about it? How would we design the doors and windows? Could we use layers instead of sophisticated technologies – for example, lobbies (with an external and an internal door) and double and triple window systems instead of highly insulated doors and windows with sealed triple glazing units? These solutions may take up more space compared to the more common solutions; however, with careful design these low-tech solutions may present more sustainable solutions compared to the more complex ones.

Doors and lobbies

Most houses have a door in the external wall that leads directly into the building, for example into a reception room or hallway. In many cases there is no internal door to create a lobby, as is common in larger commercial buildings. Opening the external door allows the external air to enter our case study house. In the winter, the colder outside air is not welcome, and results in the loss of the warm internal air. This may trigger the heating to come on if the door is left open for more than a few minutes. In the summer, the warmer outside air is equally unwelcome because it will add to the internal air temperature, influencing our thermal comfort. The door is also a means of unwanted entry, and typically it will remain closed until it is necessary to leave or enter the building. A lobby offers more scope for controlling the internal environment without necessarily compromising security.

A lobby is a space created by two doors, one opening to the external environment and one opening to the internal space. The lobby forms a space that moderates between outside and inside environments. The lobby creates a barrier to strong winds and creates a space that prevents unnecessary heat loss from the interior by closing the internal door before opening the external door, and vice versa. In very hot weather the lobby helps to prevent the external hot air from entering the cooler interior. In housing design the lobby is sometimes referred to as a porch when the lobby is proud of the main building elevation. This may have been done as part of the original house design or it may have been added later to help reduce heat loss (and heat gain), improve security and improve the visual appeal of the house.

The design of lobbies needs to consider the building orientation. They can, for example, be designed with glazed doors to benefit from solar gain in the winter. In this case, once the air in the lobby has warmed, it may be possible to open the inner lobby door to benefit from the heat without compromising security because the outer door remains closed and secure.

Windows

Most houses have one window component in the opening. This has developed from a single glazed window to windows with factory fitted sealed double or triple glazed units. While these units help to reduce heat transfer via the external fabric, they are expensive to produce, and the sealed glazing unit will need to be replaced when the internal seals eventually fail (indicated by misting of the glass within the unit). These high-tech solutions are not as durable compared to a single glazed window. Thus the question that needs to be asked is whether it is possible to design a window system that relies on two or more windows with single glazing that are mounted in the wall opening to create cavities between them. This creates what is known as a double or triple window system. It relies on a series of independent vertical layers, single glazing and air spaces, which provides an effective thermal and acoustic barrier between the inside of the building and the outside. It also allows for a greater degree of control over ventilation as each window can be adjusted to allow the required level of ventilation without creating unwanted drafts or excessive loss of heat from inside to outside. As external walls have become thicker to improve thermal insulation, it has provided the depth required to accommodate two or three windows within the opening as a possible solution to the performance challenge.

In the UK there is no tradition of having a double or triple window system, which is common in some climates with warmer and colder weather. If we consider our case study house, there is no reason why such a low-tech solution would not provide the necessary thermal insulation and a more durable and user-friendly window solution. To this we can add external and internal shutters to provide privacy and shading to prevent overheating in the summer months. The shutters also provide added security to the window. In areas prone to crime it may be that bars can be added to the windows to create an architectural feature, while also preventing unwanted access through the window while it is open. These are not new solutions; they have been used in countries with warmer climates than the UK for a long time and form part of the vernacular architecture for many countries surrounding, for example, the Mediterranean Sea. Newer interpretations of the tradition include electric steel shutters built into window and door assemblies for added security and to provide essential shading in the summer. As the UK climate changes and warms, we could look to our near neighbours to inform our choices.

Maintenance and recovery of materials

Taking a simpler approach may require additional space compared to the more complex and familiar solutions and it may require additional materials, for example two doors instead of one, two or three window frames instead of one and thicker external walls. However, there are savings to be made in the longer term because the single sheet of glass will not need to be replaced. The double or triple glazed unit will require a new sealed glazing unit once the seal degrades, which may be somewhere around 10 years after installation. Over the life of our case study house, which may be 75 years or so, the simpler approach makes sense in terms of lower expense, less inconvenience and a lower environmental impact. There may be other benefits too, such as the residents' ability to better control the internal environment to control airflow, moderate thermal comfort, and eliminate unwanted external sound. This does, of course, require the residents to understand how to operate the door lobby effectively and how to operate the windows and associated shading devices effectively. The alternative is to rely on smart building management systems to assist the residents with achieving the correct level of airflow, etc.

Reflective exercises

A conscious decision has been taken to not include any drawings, sections and details, to allow readers to develop their own conceptual details in relation to the hole in the wall, and door and window design. Continuing with our case study building, how would the details look if we embraced some of the thinking outlined in Section 16.3?

Exercise 1. Try drawing a section of a door or window design from first principles without relying on existing details to be found in the construction technology books. Think creatively; do not be bounded by convention.

Exercise 2. If Exercise 1 proves to be too challenging, try taking an existing detail, for example one of those provided in this chapter or in *Barry's Introduction to Construction of Buildings* as your starting point. Then analyse the chosen detail and explore what could be changed to create a design for the hole in the wall, in relation to a doorway or a window opening, with better environmental credentials. How would you justify your decisions?

Index

Barry's Construction Technology: Analysis and Choice, Third Edition. Stephen Emmitt and Tony Bryan.